THE GLOBAL
CLASSROOM

T0112041

THE GLOBAL CLASSROOM

HOW VIPKID TRANSFORMED ONLINE LEARNING

BY LILY JONES

Preface by Dr. Kai-Fu Lee
Foreword by Dr. Jun Liu
Epilogue by Cindy Mi, Founder and CEO of VIPKid

Skyhorse Publishing

Skyhorse Publishing books may be purchased in bulk at special discounts for sales promotion, corporate gifts, fund-raising, or educational purposes. Special editions can also be created to specifications. For details, contact the Special Sales Department, Skyhorse Publishing, 307 West 36th Street, 11th Floor, New York, NY 10018 or info@skyhorsepublishing.com.

Skyhorse® and Skyhorse Publishing® are registered trademarks of Skyhorse Publishing, Inc.®, a Delaware corporation.

Visit our website at www.skyhorsepublishing.com.

10 9 8 7 6 5 4 3 2 1

Library of Congress Cataloging-in-Publication Data is available on file.

Cover photo credit: Michael O'Neal

ISBN: 978-1-5107-5353-2
Ebook ISBN: 978-1-5107-5354-9

Printed in the United States of America

TABLE OF CONTENTS

Publisher's Note

Please note that the original content of this book was originally published in both Chinese and English in 2019.

PREFACE

The Dream of Growing Love

For many people at Sinovation Ventures,[1] VIPKid is their baby. Many Sinovation Ventures employees, including our venture partner Zhang Lijun and chief financial officer Li Puyu, have their own children studying English with VIPKid. They often share with me videos of their children taking VIPKid classes, so I seldom miss the news about VIPKid when I work with these parents, who are fans of online learning. For example, what new features did VIPKid recently launch? Even without Cindy telling me personally, I still learn about the updates early on my WeChat Moments.

---

1 Sinovation Ventures, founded by Dr. Kai-Fu Lee in September 2009, is an investment institution and platform dedicated to early stage investment and providing all-round entrepreneurship cultivation, aiming at cultivating innovative talents and a new generation of high-tech enterprises.

In the past five years, VIPKid has grown from a small team at Sinovation Ventures to an online education brand with the fastest growth rate, the largest number of students, and the best reputation in the world. It has attracted what seems like half of the Chinese investment circle, and its product reputation and revenue scale as well have far exceeded my expectations. It can be said that VIPKid is the ideal star company for many investors.

I am honored to be the earliest investor of VIPKid, having watched VIPKid grow from a small seed to a towering tree. Its growth stories are always full of love and have touched us greatly.

I met Cindy for the first time in 2013, when she had neither funds nor teams, and only a business plan. I remember Cindy came to Sinovation Ventures in a suit that day. Some people even teased her, saying that she was different from regular tech entrepreneurs. Later, when she introduced herself, I found out that this young woman was really not simple. Young as she was, Cindy had taught English for nearly twenty years!

Cindy's initial intention was to find the best teachers in the world through the Internet to help Chinese children learn more effectively. This was not simply to help them learn a language well, but also to help them know the world better through the study of a language.

Her wonderful vision for the future of education reminded me of my childhood. When I was a child in Taiwan, China, teachers taught strictly rather than encouragingly. When I was eleven years old, my family decided to send me to study in the United States. Although my mother was reluctant to part with me, she was open-minded enough to support me to travel far away and experience a new way of education.

As an international student, I had received a totally different education, one based in positive motivation from American teachers, and I gained self-confidence. More important, I learned how to deal with problems— not to seek standard answers within a predetermined framework, but to keep an open mind and accept different answers. These habits have benefited me a lot and changed my life trajectory.

Therefore, I encouraged Cindy: "Cindy, go ahead! Use your personalized and international educational philosophy to help more children fall in love with learning and try your best to change the future of education!" VIPKid officially launched at Sinovation Ventures, where Cindy built teams, recruited students, and made educational products. She was busy every day and always full of vitality.

I often observe the differences in thinking between Chinese entrepreneurs and their American counterparts in Silicon Valley. Internet companies of Silicon Valley often adopt a "going light" model. They believe that the function played by the Internet is to eliminate information asymmetry, so they always treat the Internet as a tool to connect information. By contrast, Cindy is a typical

Chinese entrepreneur who is willing to settle down and do things. In order to achieve the best educational quality, she does not mind the "going heavy" model.

Cindy wanted to build more than just a platform to connect teachers. She tried to go deep into every detail of education. She spent a lot of time communicating with parents and teachers every day to solve all kinds of problems encountered in children's learning and to find personalized learning methods and paths for each child. Later, VIPKid set up a professional service team of Learning Partners that has thousands of staff members to this day. This is something hard to imagine for Silicon Valley entrepreneurs who would rather solve problems by writing elegant code.

Cindy's decision puzzled not only Silicon Valley entrepreneurs, but also many people in China—why don't you use ready-made teaching materials rather than setting up a teaching and research team? Why not develop the market quickly to expand the enrollment scale instead of spending one and a half years polishing the educational products?

Despite these questions, I have always supported Cindy's decision and encouraged her to hold on. We can all see her persistence in education. As mentioned, the children of many employees at Sinovation Ventures were students in VIPKid's experimental class. Before moving out of Sinovation Ventures, VIPKid was the most popular topic in the workplace. Everyone liked chatting about the progress their children made in studying and the sparks of inspiration they were experiencing. Everyone passing by would be deeply affected by the energy and youthfulness of these growing dream practitioners.

The two years when VIPKid grew up rapidly were also the two years when I fought against disease. During that time, I constantly reflected on how much time I spent in my life pursuing ostentatious splendor and achievements but ignoring the really important things. Now, when I decide to do something, I will be more careful by asking myself a question: Is this something I really love? If I do it with more people, will the world become somewhat better?

When Cindy and I met not long ago, she took out her mobile phone and showed me a video. She said, "You must watch our latest video about our Rural Education Project, Mr. Lee." VIPKid organized a public welfare program in 2017 that enables teachers in North America to teach English classes in the Chinese countryside, and Cindy loves to share the stories of rural children with everyone she meets. She is a very empathetic person. Even after watching the video dozens, or even hundreds, of times, her eyes still sparkle with tears of joy.

With the Rural Education Project, VIPKid breaks geographical restrictions, allowing international teachers access to one thousand schools in the vast rural areas of China, providing regular weekly classes for more than thirty thousand rural children. These classes not only give rural children the opportunity to rekindle their interest in English learning under the guidance of foreign teachers, but also give these children in the mountains a pair of wings. Cindy always says, "Education is a window through which everyone can see the distance." This is how VIPKid helps more children open the window to the world.

Online classes are like magic, bringing me endless surprises. I was filled with joy when I saw the children from the mountain area dressed in costumes shout to the camera, "We like English class." I am proud that I set up Sinovation Ventures in 2009, which helped young Chinese entrepreneurs like Cindy hatch innovative projects like VIPKid and bring smiles and self-confidence to children all over China.

People often ask me, "Will artificial intelligence replace teachers one day in the future, Mr. Lee?" I respond that artificial intelligence can replace human beings to complete those repetitive tasks, but it cannot learn to love like human beings. Love is human nature and the most important element in education. Work with love will never be replaced by artificial intelligence.

In this book, I hope you can find not only the commercial success of VIPKid, but also the inspiring stories of ordinary teachers who joined VIPKid with their love for family, education, and the world. Here, they have forged a profound friendship with Chinese children in class. Technology will not change the essence of education, but through technology, we can bring together more love, eliminate misunderstanding and prejudice, help more people in need, and make the world a better place.

Dr. Kai-Fu Lee
Founder of Sinovation Ventures

FOREWORD

As someone who has worked in the field of English language education for more than three decades in both China and in the United States, including as the first nonnative English-speaking president for the world's largest association of English Language Teaching (TESOL International, Inc.), I am always looking for "outside the box" ideas. When I first heard of VIPKid, a company that had been making revolutionary strides in challenging conventional classroom teaching, I was fascinated and curious to learn more about the company. Fortunately, I had an opportunity to meet its Founder and CEO, Cindy Mi, on a crisp winter morning at Peking University in late 2017.

Wearing a baseball cap and casual outfit, a young uplifting woman full of energy and enthusiasm appeared in front of me when I arrived at the breakfast venue. She greeted me with a warm handshake, "Dr. Liu, I am Cindy Mi. What a pleasure to finally meet you!" In the first ten minutes, she gave me a quick overview of what she is doing and where she wants to go. I listened and asked questions. All of my questions generated even more excitement in her. I could see my questions stimulated her thinking, and she was so passionate speaking about her vision and dream with inspiration and determination that she did not even have a single bite of food. I politely reminded her to eat, but her charming smile indicated that I had already given her enough food for thought. When we parted, she said to me, "Dr. Liu, I want VIPKid to be the global K-12 online education leader in the near future. Do you think I am too ambitious?"

I smiled and nodded without a word. I had never heard a vision so far-reaching that anyone had dared to dream. But Cindy Mi did and is continuing to do so.

In fact, I pondered on this idea for more than a year, and I started reading and learning about VIPKid in every way possible, as I felt I owed Cindy an answer. When she dropped out of high school, she learned English through self-teaching, studied college curricula, and then began teaching English to kids, eventually starting her own company with great success. This was no easy feat. When she leveraged resources and found like-minded entrepreneurs to join her to launch the new business, when she assembled talented people around her from all over the world, and when she succeeded in attracting more than 70,000 teachers who are based in North America and more than half a million young learners in China (and globally) by using an online platform and innovative curriculum, it meant something big. It meant something deeper than the numbers, more meaningful than the teaching and learning, something that is shaking the conventional classroom practice, something that makes the impossible...possible.

VIPKid started to hit all the right buttons at an unusual speed: working with TESOL International to provide resources to teachers; forming an online teacher community; designing its own major course; collaborating with Houghton Mifflin Harcourt and Oxford University Press, among others, to meet the market-driven needs of learners and their parents; and developing its own assessment mechanism, with a focus on individualized learning. VIPKid was making classrooms without borders—something that researchers and practitioners could never have done within their own confined universities or schools.

In my mind, VIPKid stands for **V**ision, **I**magination, **P**erspiration, **K**ick, **I**nnovation, and **D**etermination.

Everyone can dream, but a dream without vision is an empty dream. It's better to look where you are going than to see where you have been. Four decades ago, I decided to major in English due to my family's influence. My father is an English school teacher who taught me English at home when I was a child. In my early childhood, I benefited from leafing through English books my father stored during his college years. Even though I did not quite understand William Shakespeare, John Milton, and Charles Dickens, I was fascinated by the feeling of reading. I began to be acquainted with the names of Lord Byron, Percy Shelley, Nathaniel Hawthorne, and Jack London.

One day in a family conversation, a few of these names in *Chinglish* slipped out of my mouth, and my father was genuinely surprised. I noticed that my father's face lit up. He encouraged my sister and me to start reading "Rip van Winkle" from Washington Irving's *Sketch Book* word by word, and I remember marking all the pages with Chinese translations and semi-International phonetic symbols only I could understand. It was indeed a challenge, as it was so different from what my sister and I were taught through the radio or in school curriculum. My father was very patient and used to tell us that even though the story was a challenge for us to read, once we understood it and committed it to memory, our school English would become much easier.

Upon college graduation, I became a college English teacher, and I dreamed of becoming one of the best English teachers in the field. I learned English the hard way, and I believed in better ways of learning and teaching English. Years later, I received the Excellent Teacher of the Year Award (1999) from TESOL, I started teaching linguistics to American college students (1998), and I served as the head of the English department at the University of Arizona (2007–2011). When I was elected as the President of TESOL International, Inc. (2006–2008), my dream of becoming a leader in the field came true. I believe that faith is the daring of the soul to go farther than it can see. Any human being can see farther than he can reach, but this doesn't mean he should quit reaching. VIPKid has a vision because of the faith in Cindy, and I know from my own experience that this vision will be realized.

Imagine the world we live in and anticipate the changes taking place every day. If we only follow what is to happen, we will never get ahead. When I came to the United States in the early 90s, many of my friends and relatives persuaded me to switch my major (English education) to business or law so that I could easily find a job upon graduation. I took their advice seriously and spent some time imagining my future. What I finally decided was quite opposite to their thinking. I took the unconventional route and imagined my career trajectory would be far more satisfactory by staying in language education. It is the process of learning, the experiences you accumulate, and lessons you learn that will enrich your imagination rooted in your passion. My passion is in education, and my doctorate is in education. But I minored in drama for my PhD, and I feel immensely gratified to use drama to enrich my teaching. I specialized in teaching English to speakers of other languages, but I have broadened my scope of expertise and transferred my skills to teaching Chinese to speakers of other languages. I found my expertise in language education particularly rewarding when I began to oversee international education as a senior university administrator, as I have deep understanding of where our international students are coming from, what their needs are, and how we can help them become more successful. We must reimagine the future, and reimagine the international role of English in the decades ahead in virtually every field. If we can tap into conventional wisdom with unconventional approaches, our students will be better equipped with English competence that will enable them to become global citizens. Learning English is a means rather than an end, and a high level of communicative competence is the gateway to success. VIPKid has created this opportunity and is moving on with exceeding speed to accelerate learning efficiency for the digital generation. That is the world we all imagine.

Success is always coupled with perspiration and hard work. Recalling my own career development from language learner to language education leader, I admit I have put my heart and soul into my work with a vision. One episode of my life is my three-year TESOL presidency. Among all responsibilities as TESOL president, my focus was to shift the North America-centric association to a truly international association in every way possible. From Dubai to Mexico, from Turkey to Greece, from the Philippines to Cambodia, from South Korea to Brazil, from France to Thailand, I crisscrossed the globe to advocate for TESOL's global presence. I enjoyed meeting with future "TESOLers" around the world, and I enjoyed speaking at affiliates' conferences, and in particular, I enjoyed bringing TESOL to the international communities. As a result, we have expanded TESOL Inc. to TESOL International Inc.; we increased TESOL global members and empowered TESOL professionals around the world. Yes, once a vision is set, strategic planning through perspiration is a must. I learned from Cindy that VIPKid has the hardest-working team. The book you are going to read will document creative thinking, scope of work, evidence of success, enthusiasm of teachers, satisfaction of parents, and passion of individuals representing the spirit of the company culture and its team.

Out of my professional curiosity and commitment, I had opportunities to learn more about the company culture and many of VIPKid's leading team members. I told Cindy later: "Oh, my, VIPKid has a kick to it." What I meant is that VIPKid is a unique animal. Its ambition and aspiration, its scope and scale, its speed and demand, and its rapid adjustment and changes are all unbelievable! Perhaps I have worked in US higher education for too long. What takes months to make a decision in academia could be done over one week or even one day at VIPKid. There is no time to wait, and there is no fear of change. This is happening in China. Recalling my own experience in China back in early 2005, I was invited to help design and establish an English Language Center at Shantou University supported by one of the leading entrepreneurs in Asia, Sir Li Ka-shing. I soon learned that all college students in China at that time were required to pass CET 4, an English Proficiency Test for employment upon graduation, so all students were expected to take courses for preparation of the exam. Upon careful thinking, I made an executive decision to move away from teaching CET 4 and instead help students build communicative competence as a priority. Long lines of students waited outside my office to have a chance to persuade me to change my policy. I did not. I feel that once students develop high competence in communication, they can easily pass any type of examinations. Two years later when these students took CET 4 without preparation, the passing rate was 97%. Those who had complained before ended up bringing flowers to my office, smiling from ear to ear. A few students told me, "Your approach to language education really has a kick to it." For years, Chinese students in K-12 education have been obsessed with examinations. Societal pressure coupled with parental expectations have taken the fun part out of learning. VIPKid, from its curriculum development to its course design, has revolutionized the pedagogical orientation and marked the return of a playful way to learn. That's why VIPKid has a kick to it.

Yes, an online platform, songs, gaming features, interactive teaching, and self-paced learning are just a few of the pedagogical solutions VIPKid is using—and this is innovation! Challenging the conventional way of learning, lifting up children's imagination, making learning fun, and compensating for what is missing in the school curriculum—all are in preparation for creating global citizens for the future by cultivating intercultural communication, creative thinking, and whole person learning.

In my current job as vice president and vice provost for Global Affairs at Stony Brook University, I am also responsible for recruiting students from China. Every year, there are hundreds of thousands of high school students applying to US universities. Most students will take TOEFL (iBT) or IELTS to meet the admission requirement. In China, many bright and smart students are seeking the college entrance examination routes. Even though they might be shy of English, they have solid knowledge and skills in many school subjects. How can US universities create opportunities for these students to come, as well? One of the innovative solutions is to expand the admissions channels and recruit these talented students by using their college entrance examination scores together with oral English interviews. We did it for two years by providing conditional admission as a pathway to allow these students, upon admission, to come to our global summer institute to beef up their English. It worked wonders. Those who come through college entrance exams performed better in subject courses once they matriculated, and this innovation helped those who could not easily come to a US university gain access to realize their dreams.

In a similar vein, VIPKid's educational approach has

helped many children in rural areas where the teaching facilities are poor and teachers are lacking to have the first chance to gain access to learning English with teachers based in North America. Isn't this access empowered by innovation?

Innovation requires determination. In Cindy Mi, I can see her determination and that of VIPKid. The annual facts and statistics in this book speak volumes of her determination. Once you have a vision, you need a strategic plan. Once the plan is set, all it takes is determination that includes careful implementation, execution, trouble-shooting, and, more important, determination. In the fast-paced world we live in, we are facing challenges all the time. It takes determination to deal with emerging challenges with confidence and a positive attitude.

I have faced many challenges in my career. One of the challenges is that I am Chinese and working in English education in US. I spent my first thirty years in China; how can I be accepted by US culture, and how can I emerge as a leader of universities and professional associations? The fact that I became the first nonnative English-speaking president of TESOL, and the fact that I served as head of a large English department of a research university, and the fact that I emerged as a university leader in my role as vice president and vice provost all reflect my determination to seek excellence and also to realize areas where I can improve. In a way, determination starts with knowing our own weaknesses, and then finding ways to overcome those weaknesses. We are living in a complex and changing world, and it is the complexity we live in that makes our determination durable and worthwhile. I am confident that VIPKid is facing a lot of competition and challenges on a daily basis. But determination to be the winner is carved into the company's DNA, and it will thrive and go farther and faster.

During my presidential year in TESOL in 2007, I was invited to give a plenary at ThaiTESOL, one of the TESOL Affiliates. A reporter with the *Bangkok Post* came up to me afterward and asked me, "What is the most critical challenge in learning English for Asian students?" "The lack of an English environment," I said. Without hesitation, he asked me a follow-up question: "And what is the most effective way to overcome this challenge?" I said, "Authentic input and interaction." I immediately noticed a frown on his face. I wish I could have told him what would happen in the foreseeable future: having teachers based in North America interact with Asian students online! The future is now!

By writing this foreword through my own reflection, I hope I have conveyed one message loud and clear: for the most part, extraordinary people, teams, associations, and companies are simply ordinary people doing extraordinary things that matter to them. For VIPKid to move on to the next level of excellence, it needs all six elements in its name: **V**ision, **I**magination, **P**erspiration, **K**ick, **I**nnovation, and **D**etermination. Under Cindy's leadership, I believe each and every one of these elements will contribute to helping children become global citizens.

Although the success of VIPKid can easily be defined as achieving its strategic goals, there's a difference between temporary and lasting success. VIPKid won't achieve lasting success unless another ingredient is added to the mixture, and that is to serve a cause greater than ourselves. That's what the lasting success of VIPKid is all about.

Dr. Jun Liu
Past president of TESOL International Inc.
Professor of Linguistics, Stony Brook University

INTRODUCTION

E-commerce has changed the world. In 2018, over 10% of purchases globally were made online, and the percentage is expected to keep increasing. With the rise of the Internet, companies are able to hire the best of the best from all over the world. This increase in remote work has opened up possibilities for both companies and employees. But this change is less prevalent in one sector: education. Generally, teachers are recruited to work in their local communities, and students have to make do with the educators that are available to them.

When VIPKid came onto the scene, that all changed. With its online teaching model, teachers in North America were able to expand their impact to work with children in China. Even more beneficial, the children in China gained access to experienced English teachers. The company spurred a movement of defining what teaching and learning could look like in the twenty-first century.

As of the end of 2019, over 700,000 students in China and around the world work with English teachers and learn a new language in VIPKid's engaging virtual classroom. Education is no longer confined to schools or traditional classrooms. The whole world has become a limitless classroom.

At the forefront of this change are dedicated teachers. In the early morning or late night hours, teachers around the globe go to their virtual classrooms to teach English to children in China and across the world. Many of these classrooms are lovingly constructed in the teachers' homes, where they have set up lively classrooms with posters and learning materials. These classrooms can be accessed any time, giving teachers the convenience to work extra hours and make connections with students and families anywhere in the world.

In their VIPKid classes, teachers are able to build meaningful relationships with their students and families, building global connections that will benefit generations to come. In addition to students learning English, both teachers and students get to learn about each other's cultures and backgrounds. On the surface, VIPKid is a company that delivers language classes. But it is actually so much more: it is an opportunity to learn about different people, to build cultural awareness and respect, and to empower teachers and families to take control of learning.

Students and families come to VIPKid for language learning, but they stay because of the progress and connections they make. As a class taught outside of the school day, children are engaged in learning in a fun and effective way. By making connections with teachers, students are inspired to learn in a new environment. The importance of that new environment cannot be understated. Because VIPKid classes are unlike what they experience at school, kids are inspired to participate and learn English with their teachers.

VIPKid has redefined what teaching and learning can look like. While virtual teaching is not a new concept, it had never before been done on such a large scale and with such success. The model inspires both teachers and students to engage in learning by giving them flexibility and access to high-quality curriculum that follows a clear scope and sequence to deliver impressive results.

Due to its success, VIPKid has been able to transfer its original English-to-Chinese-students model to a Chinese-to-English-students model and has begun expanding to other countries including Korea. This subsidiary of the company, called Lingo Bus, launched in 2017 and has been able to deliver a breath of fresh air to North American families looking to encourage their children to become multilingual. It is just the beginning. VIPKid is reimagining how to bring high-quality education to *all* students, regardless of their location or circumstances.

Online learning is expected to experience explosive growth, with the market share increasing year by year. According to an iResearch report, K-12 online education has become an urgent need of many parents and students. From 2013 to 2016, the market growth rate of China's K-12 online education industry was at around 30%, but it rose sharply to 51.8% in 2017. It is estimated that by 2022, the online global education market share will exceed the offline market share. At that time, the penetration rate, or percentage of a relevant population that has purchased an online learning product, will exceed 50%.

VIPKid is poised to ride this wave of online learning and change the face of global education. Its vision is to build a platform for the Internet era that inspires and empowers every child for the future. VIPKid now has offices, students, and teachers all over the world that are bringing people together all in the name of learning. In order to understand the true impacts of the company, let's uncover how VIPKid became the staggeringly successful company that it is, examine the impact its model has made on teachers and students, and peek into what's coming next for VIPKid.

THE GLOBAL CLASSROOM

CHAPTER 1
THE VIPKID VISION

In 2000, Cindy Mi was a seventeen-year-old high school student, unimpressed by school. She had just moved to Harbin, a city in Northeast China, and was behind her classmates academically. Often in classes with sixty students, she felt like her teachers didn't know her. She was struggling and no one was able to give her the support she needed, so she eventually stopped trying.

"One day, my math teacher caught me reading a science-fiction magazine during class. She tore it up, threw it in my face, and kicked me out of the classroom," said Mi.

At that moment, Mi realized that classroom learning was not for her. Though she eventually went back to that math class, she officially dropped out of high school in eleventh grade. Mi began to realize the power of engaged learning—in order for students to truly learn, they need to be cared about and feel like what they are learning is meaningful. Mi was confronted with the limitations of traditional classrooms, seeing how they are not designed to give individual attention to each student. She set out in search of a personalized education model that would inspire children to fall in love with learning.

"Curiosity and confidence are so precious," said Mi, explaining her philosophy on education.

Dropping out of high school didn't stop Mi from success. She taught herself English, completing her undergraduate education through self-study, and found her passion in teaching other people the language. She went on to become an English teacher and cofounded a brick-and-mortar English teaching company, ABC English, with her uncle. She performed a wide variety of roles within the company, starting with responsibilities such as driving and passing out leaflets to promote the company's services. Over time, the company built an excellent reputation in Beijing.

But Cindy believed that there was a better way to meet students' needs. After twenty years of teaching English, she still thought back to her own experience in high school and reflected more on what exactly didn't work. While working toward an MBA from the Cheung Kong Graduate School of Business (CKGSB), Mi tried to study online education models. She was encouraged by Professor Liu Jing and began to dream about entrepreneurship. Mi began thinking about leaving ABC English and starting her own company, but her family did not support her. Mi went to the desert to consider her options. She came back with a clearer vision for how she wanted to move forward.

As Mi learned more about teaching, she realized the power of personalization. In order to be able to truly harness student curiosity, teachers need to know their students and give them individual support. Mi reflected on what her school life could have been like had her teachers been able to connect with her and tailor learning to her interests. She even started to empathize with her former math teacher, realizing how hard it was to personalize education with so many students in the classroom. Mi began to reimagine what education could look like. Rather than having one teacher for many students, Mi imagined a world where students could work one-on-one or in small groups with a teacher that was handpicked just for them. By being able to choose teachers, students could find the best fit for themselves.

Mi thought about how she could make English learning more personalized for students. As a seasoned teacher, Mi knew firsthand the demand that China had for high-quality English teachers. More Chinese people were going abroad and studying in other countries, further increasing the need for learning English.

Parents in China were spending $15 billion per year on children's English language learning, but there were only about 27,000 qualified teachers in all of China. This was not nearly enough to meet the demands of parents, who were welcoming 16 to 17 million new babies a year in China. Chinese parents valued high-quality education that focused on cultivating abilities, and they were willing to pay for it.

Through her personal experience and her time as a teacher, Mi understood the value in learning English. She saw it as a way to build global citizens and connect to the world. The need for qualified English teachers was also clear, but Mi initially wasn't sure how to find teachers who were able to engage students in personalized and effective learning.

"How can we find the best teachers?" Mi remembers wondering.

Knowing that finding qualified English teachers in China was hard, Mi was inspired to use technology to meet this need. Mi decided to leverage the power of the Internet and the expertise of teachers overseas to meet this great demand. From this vision of an online, global classroom, VIPKid was born.

Mi brought her friend Jessie Chen, who had been working at a technology company, on board as VIPKid's cofounder, forming the company in 2013. Chen had a three-year-old son at the time and was excited about building an educational product.

"My experience was in technology, but I felt passionate about education," said Chen. "I met with Cindy and heard her vision about online education. I really loved that mission and believed we could do more together to help children learn."

At the beginning, Mi and Chen took it slow. They knew that in order to create an impactful product, they needed to get it right. So for a year, they developed and piloted curriculum that could be used in VIPKid classrooms. After a year of strategizing, Mi and Chen formally launched VIPKid in 2014.

From the beginning, VIPKid has focused on personalized learning and finding the best teachers for the program. Teachers must have a minimum of one year of experience working with children and a four year bachelor's degree. Many teachers are credentialed teachers in North America. By having a pool of highly qualified teachers, VIPKid set the foundation for creating success. When committed and talented teachers are teaching on the platform, children at VIPKid are more likely to experience results and engage in the program.

"We wanted to create a global classroom that would benefit both teachers and children," said Mi.

Another way that VIPKid creates an effective and personalized education experience is by allowing parents to choose the best teachers for their children. This is starkly different from a traditional school, where teachers are assigned to classes without much consideration of individual students' needs. By adding another layer of personalization, VIPKid invites parents to be partners in their children's education. There is not only a wide range of talented teachers to choose from on the platform, but parents are able to rate and choose the right teacher for their particular child.

The vision of VIPKid has grown beyond simply a personalized learning journey. The company hopes to change the face of global education, connecting high-quality teachers and bringing curriculum to students around the world. When kids are given the gift of language, they can truly explore the planet. VIPKid hopes to inspire and empower every child to engage in learning and harness their natural curiosity to become global citizens.

"A global classroom is critical for children of the future," said Mi.

In order to accomplish this goal, students need not just language, but also initiative to change the world. With this mission, the company is thinking not only about increasing personalization, but also creating authentic learning opportunities for students to learn real-world skills across subject areas.

Teachers began teaching on the VIPKid platform because they felt like they could focus on teaching, not the administrative tasks that often weighed them down in the classroom. Brick-and-mortar classroom teachers began teaching on the VIPKid platform to earn extra money before their work day began, but they continued teaching for much more than the money. Teachers returned to the VIPKid platform again and again.

As the VIPKid vision grew to expand both students *and* teachers, the company continued to focus on the dream of improving global education. In order to do that, the company needed to expand to meet the needs of students who would not typically have the means to join VIPKid. To do this, the company began the Rural Education Project, which brings teachers to classrooms in rural China through the digital platform, making English language learning accessible to typically under-served students.

Throughout it all, VIPKid has remained focused on developing the best product possible. It has invested in mobile technology and developed its own platform on which classes take place. VIPKid is unique because of its extended efforts to create its own video conferencing platform that works well on mobile devices, tablets, and desktops. With a powerful product, VIPKid has been well poised to change the face of online education.

CHAPTER 2
GROWTH OF VIPKID

VIPKid has always been a company of great ambition, but it's also experienced rapid growth. By the end of 2019, the company had approximately 100,000 teachers on the platform and over 700,000 students.

It all began with just four students and one teacher. In 2014, Mi and Chen were ready to start testing the curriculum that they had developed. They had received $3 million in angel investing in 2013 and were eager to start building their business. They asked the children of friends and investors to join VIPKid for their first classes. But first, VIPKid needed to find a teacher.

In February 2014, Madli Rothla was brought in to pilot the first class. While Rothla had taught English to adults before, she had never taught young kids or in an online environment. But with support, Rothla was able to quickly adapt and enjoy her new position. With a passion for music composition, Rothla was eager to bring music into the classroom.

Chen vividly recounts that first class. "Cindy and I were standing in the corner, watching quietly," said Chen. "That day, the child's mother was not at home. When the child and grandma turned on the camera and showed their faces to the teacher, we were both excited and nervous."

Rothla proceeded to teach the class to the student, watching the magic of teaching in an immersive classroom. The child followed the teacher to say, "Hello," sing the "Letter B" song, and play some games. The technology didn't work perfectly, but the class was still a success.

"The first class I had, everyone was there. Cindy and Jessie were there, and we had set everything up. Everyone was really excited," Rothla remembered. "But then we started the class, and the technology part didn't work that well. We were really worried, thinking, 'Is it really going to work. Is the Internet really good enough to even do this thing?'"

Luckily, they figured out soon after that the problem was with the colorful background behind Rothla. It had apparently taken up all the camera's processing power. Despite the technological challenges, the first student was thrilled with the VIPKid experience. After the class, the student kept shouting, "Teacher! Teacher!" even after Rothla had turned off her camera. This inspired the team to feel confident in the power of online teaching.

The early days of testing included many hiccups and learning experiences, but the small team worked together to revise their product and approach. The small format allowed them to really analyze how students and families were responding to the VIPKid methodology. Though the original team was small, they were all committed to the success of the company. There was excitement in working through problems and making the product the best that it could be. Seeing actual students engage with VIPKid was motivating and exciting to all of the employees.

"We were all dedicated and passionate about education," said Mi. "We shared the same values and wanted to do something that made a difference in the world."

Rothla describes the early days at the company as typical start-up culture, with everyone pitching in to do a wide variety of things. She remembers the initial staff consisting of about five people: Cindy Mi, Jessie Chen, a woman in marketing, Rothla's manager, and Rothla. The company was housed in a large incubator in Beijing. The incubator provided a lot of resources that VIPKid was able to take advantage of, such as legal services, a finance team, and even food.

"It was really nice, we had these lovely ladies cooking food for us and then bringing it out in huge pots, and you would just go and scoop up some fried dishes or noodles," said Rothla. "Everyone loved noodle day. I think they had noodle day twice a week."

Rothla's job was varied and interesting. She developed and sang the "Hello Song" and "Goodbye Song" that teachers and students still sing today. She continued to teach, while also helping to improve and develop the curriculum, answer teacher requests, and communicate with teachers and students.

"You'd come in and maybe you'd answer some emails with teachers' questions, and then the next thing, you'd be making a PowerPoint for a lesson. And then the next thing, you'd be, I don't know, preparing a marketing event. And so on and so forth," said Rothla.

While the small team was working hard to grow all areas of their business, they were also finding their voice. The curriculum team worked to perfect the initial curriculum and teaching style, all at the same time that students became more accustomed to learning in an online format.

To help students and families see the value of the VIPKid model, the company began to run experimental classes, each of which was limited to ten people. In the second round of experimental classes, Mi refunded half the tuition fees to each student. In the third round, the normal market price was charged.

"Only paid users will experience it seriously," Mi said.

"It took us a year to find a hundred students," said Mi. "We started with four, then got about ten new students each month."

At the beginning, parents wondered why they should choose an online school instead of in-person classes. Mi was also having trouble finding adequate financing, with nine out of ten companies rejecting VIPKid's business model because their investors would not support it. Mi decided to put effort into proving that the VIPKid model would be successful, thinking that if she could show proof of concept, investors would be more likely to get behind the company.

But through persistence and outreach, VIPKid was able to get many more children and parents on board. The company continued to test its product, signing up more teachers and students and learning from their experiences. VIPKid invited parents to talk face-to-face, do an assessment with their children, then try a sample class. Parents were able to view the class and see their children's progress and engagement. The students came eagerly to their classes and enjoyed their experiences.

Parents were happy to find that their children's English levels improved and decided to recommend the classes to their friends. VIPKid relied heavily on referrals, which allowed the company to leverage satisfied families to bring in more families wanting to try its platform. At the same time that the parent and student users were growing, the teacher base was increasing, as well. All of the original teachers were recruited by Mi, and it wasn't always easy to sell teachers on the power of the approach. In general, the teachers were hesitant to believe that they could teach Chinese children online without knowing the language themselves.

"The first twenty teachers were all recruited by me. I talked to the teachers through Skype. Basically, it took five hours to convince them, to show it was especially easy to teach online," said Mi. "One of the teachers said, 'I didn't think about teaching online. I didn't know how to teach students if they couldn't speak English, and I couldn't speak Chinese.'"

Mi addressed the teachers' concerns directly, saying, "It doesn't matter. We create the content for you and you communicate with the children with body language according to the content. For example, when talking about a puppy, you can bark. You don't have to know what type of animal the dog is. Therefore, what you teach the student is actually a kind of cognition. You can show your dog to the student."

One by one, the teachers were persuaded to begin teaching on the platform and offering their services as independent business owners.

With teachers, parents, and families coming on board in greater numbers, Mi was ready to scale the business. After showing this growing interest in VIPKid, Mi was able to secure investors. Many of these investors were parents whom Mi found to be more willing to promote VIPKid and who provided valuable feedback to the company. But finding investors for the A-round financing was not easy.

"Cindy was responsible for financing. Every day, she would visit five to six investors with a computer, needing to talk for two hours," said Chen.

Mi visited investors at this rate for several weeks, which was exhausting. One day, Chen saw Mi lying on the company floor with a backache. Chen began to cry, seeing the great effort Mi was putting into the company.

"The pressure on her is twice as great as ours," Chen remembered.

But Mi's efforts caught the attention of investor Lixiong Niu. Niu was interested in Mi's success with ABC English and the idea of using English teachers who taught from home. But he worried about delivering the lessons through one-on-one video. He believed that parents would be hesitant to sign up their students for online English classes and wasn't sure if VIPKid had enough data to prove its success. At this point, there were only forty VIPKid students.

But Niu continued to be interested in VIPKid's approach, particularly inspired by the idea of bringing high-quality English instruction to students in cities that may not have previously had the opportunity to take such classes. Then Mi brought him data about the twenty-eight students who had completed the VIPKid experimental classes. They had a 100% renewal rate.

These data convinced Niu that he must invest in the company. An A-round investment of $10 million US dollars was made. This proved to be a very good investment.

As of 2019, the valuation of VIPKid has exceeded $3.5 billion.

This confidence from investors continued. In October of 2014, VIPKid received $5 million dollars in Series A financing, followed by another $20 million in Series B fundraising the following year.

Growing Bigger and Better

VIPKid also started to appear in competitions and gain recognition for its emerging products. In 2014, VIPKid participated in an innovation competition organized in Silicon Valley, winning 7th place. The following year, the company represented China in the Shengjing Global Innovation Awards, where it placed in the top six.

In 2015, the teaching and curriculum team had grown to about twenty people. For Madli Rothla, that meant that her job became a little more specialized. She was no longer teaching. Instead of fielding hundreds of emails from teachers each day, she transitioned to writing curriculum. Having experience teaching and perfecting the product made Rothla an asset to the curriculum development process. Despite moving into a more specialized role, Rothla describes the VIPKid company culture in 2015 as a start-up environment where everyone worked on whatever needed to get done.

But Rothla came to enjoy curriculum design, saying, "It's really fun, and I didn't know I would like that so much, just sitting in front of an Excel, thinking, Which word should go with this lesson? Which kind of point is the best to teach here? But it's actually quite enjoyable."

The curriculum at VIPKid was first developed at a very basic level. It focused on teaching letters, letter names, letter sounds, and some vocabulary. As more staff members came on board and more students signed up for VIPKid, the curriculum became built out into what the company calls its "major course line."

Later, the VIPKid team moved to an office right in the middle of old town Beijing, right behind ancient buildings called the Drum Tower and the Bell Tower. The VIPKid offices were basically inside an old Taoist temple. Though employees couldn't enter the actual temple, they were in the surrounding buildings.

"If you wanted a peaceful moment, you would go walk through these little hallways and you would end up in front of this Taoist temple that was just there," said Rothla.

As Rothla was continuing to focus on her teaching and research work, the company was growing, as well. The A round of financing had proved the feasibility of the VIPKid business model. Now the biggest challenge the company faced was figuring out how to quickly scale. In the first year and a half of the business, the number of users had grown very slowly. In 2015, VIPKid was only adding forty new users per month.

After the Spring Festival in 2015, the plan was to fully introduce VIPKid to the market. But then the sales team leader unexpectedly resigned from the company because of the long hours.

With the loss of the sales team leader, Cindy Mi began to lead the sales team personally in March of 2015. In April, the number of VIPKid new registrations exceeded one hundred. Mi remembered the joy that came when the hundredth parent made a payment at 11:40 p.m. In their excitement, the entire team went out to celebrate by eating lamb skewers and having drinks. They ended up staying out until four or five in the morning.

As the company continued to grow, Victor Zhang joined as a cofounder in May of 2015. He is responsible for the company's overall market operations, sales, and services. Prior to joining VIPKid, Victor Zhang was the cofounder of Baicheng Travel Network.

Mi had spent half a year trying to convince Zhang to join the team. Initially, he wasn't sure, but Mi kept calling. Mi called Zhang every two weeks. Eventually, her efforts paid off when Zhang joined VIPKid as a cofounder. When he came onto the team, Zhang was presented with the challenge of how to scale the company. He began by looking at how to attract students. After getting this market established, he then turned his efforts to the conversion of students. After focusing on this, sales increased and service could not keep up. He then refined the service system. Following this market-sales-service cycle has been key to VIPKid's success. As Zhang noted, many traditional marketing practices didn't work with the VIPKid audience. This is because the customers for VIPKid are so specialized. Not only do they need to have children, but the children need to be a target age. Zhang set out to find out how to best reach this target market.

Instead of doing large-scale advertising, Zhang focused on opinion leaders that moms support, including some WeChat accounts. While researching this, Zhang realized that some of the opinion leaders are VIPKid users themselves. Early adopters of VIPKid were middle-class families with relatively high income levels and regular Internet use. They are also opinion leaders.

A Key Opinion Leader posted a video of a child and teacher on Weibo, a Chinese microblogging website, which sparked the curiosity of many parents. VIPKid was able to get its website and customer service phone number added to the Weibo post.

"The office was crazy. The phone was ringing," recalled Jessie Chen. "There were two thousand registered users in a week."

Zhang and the rest of the company were thrilled to see that the brand was beginning to connect with a community of parents. Once news started getting around about VIPKid, word of mouth helped the company expand its reach. VIPKid users recommended new users and got incentives when their referrals signed up for the program. In 2015, more than 50% of new users came from referrals. To this day, the acquisition cost of VIPKid users is the lowest in the industry. Because of the strength and success of its referral program, VIPKid's acquisition cost is unmatched by similar companies.

Knowing that the company was poised for expansion, Zhang set a target of 30% monthly growth. The education industry generally has a 10% to 20% monthly growth target, but VIPKid was ready to set its sights high.

The sales partners were hesitant about Zhang's lofty goal. Without everyone on board, the goal was initially not able to be reached. Zhang was determined to figure out why the goal was not met. He looked at the conversion rate, sales training, and other possible reasons. He finally concluded that the problem was psychological.

"The sales team was very resistant in their hearts. They were used to setting a goal that they could achieve, but the potential of people was better than this," Zhang said. He saw that the potential of growth was much higher than originally perceived. He just needed to get the rest of the team on board.

At the time, the founding team expected to get about two to three thousand students by the end of 2015. The current strategy was to let sales break through on their own. But since there were only a few hundred students when Zhang started, he didn't think waiting for sales in this way was enough. Instead, he motivated the sales team.

After one month, the expected sales figures were not met, but the number of students increased by 20% to

30%. This was enough to satisfy the sales people and help them to gain confidence in Zhang's approach.

In August, sales partners came to Zhang to discuss the number of sales that would happen the following month. Zhang suggested a goal 30% higher than the company was currently doing. When the sales partner didn't say anything in response to Zhang, he asked if the number was too high.

"You are underestimating us," the sales partner said. He then suggested a number one hundred more than Zhang's original goal. In the end, they exceeded the original goal.

In order to recruit more users, VIPKid launched a policy in May 2015 where users could receive a refund on their first twelve lessons, unconditionally.

"I discussed with Victor for a long time," said Mi. "Finally, everyone decided that this was risky, but it still needed to be done."

The policy proved to be worth it. Parents trusted the company more because they could receive a refund if they were unsatisfied with the product. The company found that more parents chose to give them the opportunity to improve the product and did not abuse their right to receive a refund.

This helped to get more people to try VIPKid and proved to be quite successful. By the end of 2015, the company had achieved six thousand paying users, and the refund rate was controlled at 2%.

Through this period of growth and still today, the three cofounders remain committed to working together. To do this, they have divided some of the company's duties and come together to fulfill others. Chen is responsible for the teacher side of the business, including making sure that the teacher community is successful and happy. Zhang handles the operations side of VIPKid, including marketing, sales and services, and helping Chen with the teacher community. Mi is responsible for new business, including product, content, and strategy. When dividing their responsibilities, the cofounders played to their strengths and experiences.

While keeping focused on growth, 2016 was a fruitful year for VIPKid's fundraising efforts. In April 2016, it raised 100 million dollars in its C financing round. This investment allowed VIPKid to become the largest online children's English education company.

This was also the year in which the late, great American basketball star Kobe Bryant announced the creation of Bryant Stibel, an investment firm focused on early-stage businesses. Bryant Stibel strategically invested in VIPKid in a joint investment with Learn Capital, an American education venture capital firm. VIPKid was the second Chinese company that Bryant Stibel invested in. Kobe Bryant himself said it only took him five minutes to decide to invest in VIPKid.

In 2016, a large VIPKid poster went up in Times Square, New York, which the company saw as its announcement of itself to the world. The year 2016 was also when VIPKid received an increasing number of awards. In December of 2016, Cindy Mi received recognition for her accomplishments as an entrepreneur. She won the "Win in China" competition and received the title of "Entrepreneur of 2016" after getting the highest votes. Just two years into starting the company, Mi was being recognized for her efforts toward getting VIPKid incredible results.

Expanding VIPKid's Impact

In 2017, VIPKid collaborated with the Jack Ma Foundation, becoming the first professional English company to join the education public platform under the Jack Ma Foundation. In August of 2017, VIPKid finished its D round of financing, raising 200 million dollars. This was the largest amount raised in the Global K-12 investment field. In part, this money helped to launch Lingo Bus.

In 2018, VIPKid raised $500 million in its series D+ financing. This investment made the company the only education startup in China valued at over RMB 20 billion (approximately $3.5 billion). At the end of 2018, VIPKid was valued at over $3.5 billion. In just five years, VIPKid has gone from a vision to a lucrative business.

In February 2018, *Fast Company* ranked VIPKid as the 29th most innovative company in the world and the second most innovative company in China. VIPKid also made the *Forbes China* list of China's most innovative companies in 2018.

Many attribute VIPKid's success in large part to the strength of its teacher community. Teachers' satisfaction with teaching on the VIPKid platform continues to increase. VIPKid topped the list of *FlexJobs*' 100 Top Companies with Remote Jobs in 2018, surpassing Amazon.

By the end of 2019, the VIPKid community has grown to approximately 100,000 teachers on the platform. Many attribute the company's growth to its scalability. There are many Chinese students wanting to learn English and many teachers wanting to teach with VIPKid. Because of this synchronicity, the company can expand easily as it brings on more and more teachers and students.

How Things Have Changed

The first thing that comes to mind when thinking about how much has changed for VIPKid is the size. The sheer number of students, employees, and teachers on the VIPKid platform has exploded.

For example, Madli Rothla was once the company's only teacher during the pilot stages. Just six years later, there are over seventy thousand teachers at the company. Employees have experienced this growth in a variety of ways. While some of the departments are still in the old Taoist temple that Rothla described fondly, her department moved in 2018. There are now three VIPKid offices in Beijing, as well as offices in Chengdu, Dalian, and

Shenzen. There is also an office in San Francisco, and other employees throughout the United States.

In 2019, the curriculum team has grown to more than 400 people. Rothla now belongs to the academic planning team, which is where all curriculum projects start off. "The company or the management has an idea, okay this is what we want to do, these are the kind of students that we want to target," said Rothla. "Then we will come up with a detailed design of what those ideas could look like in practice.

Reflecting on how VIPKid has changed over the years, VIPKid's Community Production Manager Kevin Gainey said, "The things we stressed about seem so small now." Gainey remembers a big pain point of the early days being that teachers would need to complete a Word document if they wanted to give students an assessment, then send it back to the company to track students' progress. If teachers weren't submitting their feedback, it was hard to gather key assessment data.

But eventually, a technology fix made this problem obsolete. Teachers can now give the assessment during class, collect the important assessment data, and press a "submit" button to immediately send the information to the company. Now there's a 99.9% submission rate. Improved technology and an expanded, more specialized team have made the company more efficient.

"I feel my goals have shifted. It used to be very, very tangible what I needed to do," says Gainey. "For example, we need teachers to be able to teach this concept this way, so students can understand it. Now the big goal is, How do we make sure teachers want to continue teaching with us? And it's a bigger goal, but at this point, it's what's most important to us because all the other stuff, the things that would make their lives more complicated or more annoying, they have been fixed."

For Rothla, the biggest change is that the way that the company is run has become more procedural. While employees wore many hats at the beginning of the company, people's individual jobs have become more specialized in recent years. Now there is a clearer line between which jobs fall to which teams.

"On one hand, I think it's good because everyone should know what they're responsible for," said Rothla. But because Rothla came into the company at the very beginning, when things were bustling and new, she also feels a change in the pace of the company. Though it can be relaxing, she sometimes misses the days where everyone had to go above and beyond on everything.

Overall, Gainey said he has not been surprised about the company's growth, because the product is of such high quality. "It makes sense because of what it has," said Gainey. He also cites the scalability of the company as a reason for its success.

Personalized Career Paths

Throughout all the changes in the company, one thing

has remained: a passion for personalization. While this is most apparent when talking about the educational approach to teaching, the focus on personalization is clear within the company, as well. When talking to employees of VIPKid, they also seem to have enjoyed a personalized career path.

Stephenie Lee, currently a product leader, has been with VIPKid since 2017. She began as the senior manager of learning solutions, working to create VIPKid curriculum and digital products, including a digital library of resources. She then became the senior manager of education product strategy, working on ways to visualize student progress and ensuring that the intent of the curriculum team in designing content would translate into a strong overall student learning experience. During that time, Lee thought about how to help kids understand that process, all while considering the company's curricular strategy.

Throughout her career trajectory, Lee has been challenged to take what she's known about the company and its customers and create the best product possible. Lee attributes much of VIPKid's success to its scalability. Once the products are created, tested, and perfected, Lee says that the key is "ensuring that our sales, our marketing, and our service team members understand the product enough to be able to help users want it, understand it, and use it effectively."

Currently, Lee is helping VIPKid scale by reaching

more kids through Massively Open Online Courses (MOOCs). When Lee joined VIPKid in 2017, it was starting to do some product diversification. VIPKid was moving from its major course line with its flagship one-on-one blended learning model to having more products. VIPKid started first with a digital library, which is a completely digital tool, with readers that you can read along with and play audio for. Then VIPKid expanded to include MOOCs, which bring one teacher to thousands of students. Currently, VIPKid offers thousands of live MOOCs at a variety of language levels.

Kevin Gainey has experienced a similarly personalized career journey at VIPKid. Kevin was an early employee with the company, joining in 2015 as a teaching quality and operations specialist. As Gainey describes the company culture when he joined, "everybody was doing everything." His initial job was a mix of working with teachers and doing operations work. Because there were only about 150 to 200 teachers teaching on the VIPKid platform at the time, each employee who was hired was given a batch of teachers and tasked with addressing the problems the teachers had. The employees took care of everything that their assigned teachers needed, from contract signing to helping with technology issues.

But as the company grew, each employee's job became more specialized. For Gainey, that meant that he went from doing all sorts of tasks, including being on the first iteration of the company's communications team. At that time, there was a single inbox that teachers would email to, and then there was a group of employees that had to answer all of their questions every day. From this experience, Gainey was able to get a better understanding of the teacher experience.

After about a year in this role, Gainey moved onto another team. Being on this team was a great extension of Gainey's previous experience, as its goal was to try to understand and provide resources to help improve teachers' experiences on the platform. Gainey's team worked on creating optional resource videos, where they would provide tips on how to teach certain concepts or parts of the VIPKid curriculum.

Through this experience, Gainey learned how to make effective videos. When making videos, he tried to appeal to different types of learning styles, similar to what teachers do. The videos had a visual component, where teachers could read information, recorded examples of teaching that teachers could watch, and examples for teachers to be physically involved. The physical component mirrors the Total Physical Response (TPR) approach that teachers will often use in their classes. In the videos that Gainey produced, participants often submit a demo class video, where they actually go through the motions of teaching a VIPKid class.

Again, his experience making videos led Gainey down a new path at VIPKid. As the company grew, there was a need for more videos and resources. As part of a digital library, Gainey was recorded reading English bedtime stories to students. Today, many students listen to Gainey reading these stories every night. Gainey has also recorded songs for the resource library and created a series of Virtual Field Trip videos. For the Virtual Field Trip project, Gaines travels around the world and records videos of himself interacting with local residents and visiting interesting landmarks.

Today, Gainey's job is split between supporting teachers and creating resources for students. His job has been able to adapt to his needs, strengths, and preferences, providing him with a personalized path. He says, "What I like about my job is interacting with people and creating meaningful content." During his time with the company, Gainey has been challenged to constantly stretch and improve. As he says, "The phrase that encompasses what we are doing is 'do better.' That is what we are doing here, we are trying to do better."

By providing employees with chances to grow and shine, VIPKid supports both the company as a whole and individual employees to do better. Looking at the career trajectories of Lee and Gainey, it is evident how personalization and creating opportunities for growth can lead to employee satisfaction and job success.

Company Culture

As a company, VIPKid runs with the goal of "love children, understand education." Throughout everything it does, the company hopes to make big impacts in the lives of children. Working together for a shared goal has helped to create a strong company culture and bring employees together.

As well as enjoying the benefits of personalization, employees at VIPKid have opportunities to connect with teachers and students. Whether at the Journey teacher conferences or hosting teachers at the office in Beijing, employees feel very connected to the entire VIPKid community.

Throughout the history of the company, employees of VIPKid also seem to have enjoyed one another's company. During VIPKid social events, it's not surprising to find popular VIPKid songs being sung. Rothla remembered the 2016 annual party, where the team went out to a small restaurant in old-town Beijing, a 1980s-themed restaurant with school desks inside. At the beginning of the party, all of the employees sang the VIPKid anthem and made hand gestures, like what students and teachers do at the beginning of a VIPKid class.

Though now the annual parties are much bigger, the enthusiasm for VIPKid remains. Each year, the curriculum team always designs a show. In 2018, Rothla and a colleague made a medley of popular animal theme songs.

Fifteen other colleagues dressed up in animal costumes and danced while they sang songs.

VIPKid makes learning fun for kids, but it also makes working fun for employees. When we speak with VIPKid staff, it is clear that they share a commitment to making the world a better place by harnessing the power of education. It's also clear, whether it is the employee parties or the dance sessions that teachers participate in during their Journey conferences, that joy is a major ingredient in VIPKid's success.

VIPKid employees have also come together to help children in a variety of ways. Once, a twelve-year-old VIPKid student named Miya came to Beijing from Southern China because of a serious illness. She had an imminent operation and urgently needed a blood transfusion, but the stock in the Beijing blood bank was not enough. Miya's parents had to find six type A blood donors, or else Miya's surgery would have to be postponed.

Miya's family did not have relatives or friends in Beijing, so they had difficulty seeking out blood donors. Eventually, Miya's father thought of the VIPKid Learning Partner who Miya had worked with. Fifty-five eligible volunteers signed up to donate blood in less than two hours once VIPKid got word of Miya's struggles. In less than two days, the entire amount of blood needed for Miya's transfusion was collected. From the time of learning about Miya's condition and the successful end of her operation, the entire process only took six days.

CHAPTER 3
THE VIPKID MODEL

VIPKid was built on the idea of a personalized, global classroom. Its success is due to a convergence of needs and desires: it caters to the busy modern life, bringing a convenient way to learn English to overbooked families. It brings the best teachers and engaging lessons to kids who need a little extra fun.

The VIPKid model is successful because it works for students and teachers and benefits from its scalability. Instead of a brick-and-mortar supplemental English learning experience, which maxes out at a certain teacher-student ratio and has size constraints, VIPKid can be used by many people at once. The growth of VIPKid largely rests on its ability to scale so efficiently.

But even with its ability to scale, there needs to be enough interested students and teachers to make the model work. Luckily, VIPKid is in no shortage of either group. Parents and students come to VIPKid because it is an effective way to teach language. They appreciate being able to be matched with high-quality English speakers and have their children complete regular classes from the comfort of their own home.

"Kids in China take different types of courses for the whole week, such as tennis, math, or piano, and most of the lessons are on site," said Rina, a parent of a VIPKid student. "Traffic is not good, so it takes a long time to bring her to her lessons. I also need to stay there waiting for her to finish the classes."

But with VIPKid, parents like Rina can be in their own homes while their children complete their English lessons. Having lessons at home also cuts down on commute time for parents, kids, and teachers, saving valuable time for busy families.

Teachers also appreciate the flexibility that comes with VIPKid, often teaching out of their home office.

Of course, the most important component of the VIPKid model is its successful learning outcomes. All of the convenience and fun are perks on top of the language learning that children engage in and the progress they make. VIPKid takes a well-researched approach to language learning and total immersion and transfers it to an online environment. This means that children experience English lessons taught entirely in English from their first day with VIPKid.

While this may seem overwhelming, it is effective. Its effectiveness is due to the thoughtful way in which

VIPKid helps students go through a scaffolded course progression. When parents enroll their students in VIPKid, the children make their way through a progression of language-learning courses that the company calls its "major course line." Putting students through this well-tested curriculum ensures that they make steady progress in learning English.

Though lessons may be entirely in English, students are supported in understanding the content through body movements and facial expressions. These components make the lessons fun and different from the education students may receive during the school day.

Engaging Students

Students appreciate the energy and enthusiasm that teachers bring to the classroom. Rina Xie is a parent whose daughter Melody took VIPKid classes for three years, making so much progress in her English language development that she is now interviewing to go to high school in the United States. After Melody's screening process with American high schools, Xie reflected on how her language lessons with VIPKid impacted Melody's performance.

"The teachers are very energetic and well prepared," said Xie. "That has affected my daughter. When she talked to the candidates, she was energetic and confident. Melody knows that when she talks to somebody, she wants to express her true self and the best of herself."

For Melody, her favorite part of VIPKid was sharing and interacting with the teachers. She was less interested in doing homework or formally practicing English, and much more interested in talking with her teachers. She enjoyed sharing about herself and learning about the teachers.

But it's not just being engaged that matters; having these authentic conversations helps students like Melody to develop fluency in English. There is no faster way to learn a language than total immersion in that language, which works best when students are active participants. For Melody, she was so excited to learn about her teachers and another culture that she would often repeat, in English, what she learned to her mother. These conversations resulted in additional authentic speaking practice.

Building Relationships

There is a novelty in having friends around the world, which results in greater engagement in VIPKid lessons. When they have strong relationships with their teachers, students are motivated to learn English for an authentic reason: to get to know and converse with someone they care about. Despite their lessons happening in a virtual format, VIPKid recognizes the immense impact strong relationships have on learning.

Time is built into every VIPKid class for relationship building and sharing about one another's culture. Part of the thrill of engaging in such an online platform is interacting with someone across the globe. Including time for teachers and students to learn about each other increases engagement and provides an opportunity to practice speaking and listening skills. The VIPKid curriculum also leverages this curiosity to include lessons based on learning about particular aspects of culture, like holidays and celebrations.

When Melody first started learning English through VIPKid, Xie would often sit next to her. But as time went on, Melody requested more privacy. As Xie found out, Melody really enjoyed talking to her teachers about her life. She would often ask to speak with one of her teachers, not always realizing that chatting about her life was also a great way to practice language skills. Sometimes Melody would even request that she meet with certain teachers to finish their conversations. These teachers became trusted confidants.

Relationships are easier to build when you interact with the same people on a regular basis. Realizing this, Rina appreciates the consistency of teachers at VIPKid. Before taking VIPKid classes, Melody took in-person English lessons with English speakers from abroad. But these teachers were usually only in China for a year or so, which resulted in inconsistency. At VIPKid, the flexible nature encourages teachers to stick around. This in turn helps teachers to build strong, long-lasting relationships with students.

With previous English teachers, Rina and Melody not only experienced teachers leaving, but also a range in the teachers' abilities and experiences. With VIPKid, Rina felt more confident in teachers' experience, as she can look at their educational and work histories on the website. Seeing the teachers' qualifications and knowing the quality of VIPKid's curriculum helped Rina to trust that Melody was engaging in an effective educational experience.

Like Melody, eleven-year-old twins Claudia and Chris appreciate the relationships that they've been able to build with their teachers. After taking classes with VIPKid for two years, the twins have built relationships with a collection of teachers.

"I talk to my teachers about being bullied," said Claudia, sharing how she feels comfortable being vulnerable with her teachers.

Claudia's relationships with teachers have helped her to not only learn language, but also navigate difficult social situations. It's amazing how having a friendly ear to share experiences can help students feel supported. Claudia's ability to open up to her teacher also showcases the strength of their relationship. Generally, students share the most difficult experiences when they feel comfortable.

"When I reunite with my teacher, I like to share something about my life before even starting the lesson. That's my favorite part," said Chris, sharing how he values getting to know his teachers. "When we start a lesson, I could start talking about Star Wars. That's how you build relationships."

Claudia and Chris found it particularly easy to get to know teachers because the teachers are often so energetic and welcoming. The culture of the classes is different from traditional Chinese schools, where teachers often are very strict. But at VIPKid, the teachers feel more like part of the family.

In addition to building strong relationships, Claudia and Chris are able to get a more intimate learning experience with VIPKid. During the regular school day, the students don't usually get a chance to work one-on-one with teachers. VIPKid lessons are more interactive and provide a much more personalized experience than the one Claudia and Chris receive at school.

"At school, the teacher sometimes doesn't notice me in class," said Claudia. "With VIPKid, I finally have a chance to work directly with a teacher. I don't have to ask them anything after class."

For students, these relationships are what gets them excited to come to class. For teachers, it's no different. Showing up for their students motivates teachers to get up early in the morning and greet their classes with positive energy. One of these teachers, Rebecca Phelps, began teaching on the VIPKid platform after working with kids for as long as she can remember. Even after so much experience with children, Phelps says she never really understood the power of relationship building until she signed up for VIPKid.

"I started to get to know them and their likes and dislikes, got to know their families, their siblings, where they're traveling, and what are their dreams," said Phelps. "It was amazing to me that relationships could be formed over the computer in a twenty-five-minute class."

Over her time with VIPKid, Phelps has gained many regular students who have grown in both their language learning and in their daily lives. She has had the pleasure of seeing students develop all sorts of new skills, like when she watches a very nervous child gain confidence and blossom. Watching a child grow is one of the most rewarding parts of teaching on the VIPKid platform.

In 2019, Phelps got the opportunity to meet a couple of her students in person, which was incredible. Phelps traveled to China on an adoption trip for her youngest son. On every adoption trip, prospective parents must go through Guangzhou, which was close to where several of her students live. Phelps shared with one of her students that she was traveling to Guangzhou. The student was very excited to realize her beloved teacher would be nearby. The student told her mother, and they decided to come to Guangzhou to meet Phelps's family and take them out to lunch. Both Phelps and her student were thrilled to meet in person.

Jessie Chen, cofounder of VIPKid, has witnessed several meetings of students and their teachers. At 2018's

Journey conference in Orlando, Florida, several students flew from China to attend the conference and meet their teachers.

"There was a moment when one teacher met with her student, whom she had taught for several years. As the teacher hugged the student closely, many audience members began to cry. It was so touching," Chen said.

Feedback

The strong relationships and engagement of VIPKid lessons help children to be successful in their learning. At the beginning of their lessons, students sometimes have their parents sit next to them and observe the lessons. But as students get older and more confident with the VIPKid experience, they often complete their lessons independently. This is great for the students but can leave parents less aware of their children's progress.

To help involve parents, teachers send reports of children's progress after every lesson. These messages help to keep parents informed about their children's learning and support them at home. The interactions between parents and teachers helps to build trust and connections between important adults in students' lives.

Julia Lee, Claudia and Chris's mother, is the VP of User Experience at VIPKid in Beijing, where she travels twice per week and spends time away from her family. While she is away, Claudia and Chris take regular VIPKid classes. Even though Lee is not physically close to her children when they take the lessons, she is able to stay informed on her Claudia and Chris' progress through teacher feedback.

"Not only do teachers write about students' progress towards the lesson objectives, they also wrote personal things about Claudia and Chris," said Lee. "They express their personal feelings about my children in their class feedback. That's so impressive."

Giving this regular feedback helps keep parents informed, but it also helps parents to gain confidence in the VIPKid platform. By staying up-to-date about what their children are learning, parents see how VIPKid is effectively teaching their children English. This can be particularly helpful for parents who do not speak English themselves and may not be able to accurately gauge their children's progress.

Teacher Choice

By being a virtual company, VIPKid has the strong advantage of being able to place Chinese students with English teachers. By learning from a teacher that lives in another country, students are able to pick up on nuances in the language and learn correct pronunciation. But VIPKid goes one step further, allowing parents to choose from a wide variety of teachers in North America. By doing this, parents are able to customize their child's experience. Not only can parents choose the best teacher for their particular student, they can also choose to use different teachers over the course of their time with VIPKid.

But VIPKid is not just based on location flexibility; it also embraces a popular flexible format. Parents can view teachers' profiles, taking their education and work experience into account when choosing the right teacher for their particular child.

Rina, a parent whose daughter Melody has been enrolled in VIPKid classes for almost three years, saw the benefit in this feature. For Melody, being able to interact with different teachers gave her the opportunity to talk with them and learn different styles of speaking.

"I intentionally had Melody work with different teachers," said Rina. "I think that effective communication means being able to speak to different people."

Rina and Melody chose to use three teachers at the same time, rotating through who Melody worked with. This gave Melody a chance to build strong relationships with each teacher (which would have been harder with more teachers) while being exposed to different teaching and speaking styles (which would have been harder with fewer teachers).

Stephenie Lee, a product leader at VIPKid, agrees with Rina, saying that having different teachers "makes sure that kids get exposed to a lot of different accents and pronunciations as they are learning English, allowing them to be able to transfer their English listening and speaking skills more effectively in a real context outside of the VIPKid classroom."

When Melody came to VIPKid, she had some knowledge of English but lacked the ability to speak in full sentences. Rina felt that Melody needed to develop the confidence to try holding a conversation in English.

VIPKid was the perfect solution for Melody, as it focuses on developing speaking and listening skills. Having structured time to speak with English speakers was extremely beneficial for Melody, who developed the skills and confidence to talk to her teachers through her regular VIPKid lessons.

Other times, parents choose to stick with one teacher for their child. This approach provides consistency and helps students to build strong bonds with one teacher. For example, Kim Fortner, a teacher, sees the same student at 8 a.m. five days a week. She is able to make great progress with that student, both in terms of language development and in building a trusting relationship. This relationship is more similar to one that regular classroom teachers, who see the same kids every day, have with their students.

"Now I have my regular students," Fortner said. "Families are able to take a trial in the beginning before they sign up to see what VIPKid is about. Many times when you give these trials, you develop your regulars from this. You're the first impression they had and the parents are so excited, so they connect with you and become your regular."

Whether families decide to go with one regular teacher or rotate through different options, VIPKid gives them the flexibility to do what is right for them. This is just one more layer of personalization that students and families are offered through their child's journey of learning English.

Lesson Time

At beginning and intermediate levels of VIPKid classes, instruction lasts for twenty-five minutes each class. Rina initially was hesitant about this length of time, wondering if it would be enough time for effective teaching. But she was pleasantly surprised with how the classes held her daughter Melody's attention.

"During that time, they can fully participate and be active," said Rina. Having short sessions twice or three times a week is a consistent way for students to learn English."

Melody spent three years with VIPKid, moving through the different levels. Over the course of her time with the company, Melody has become confident. In addition to becoming more fluent in speaking English, Melody has learned more about other cultures through interacting with teachers. In fact, Melody is currently applying to high schools in the United States. During her interviews, Melody was often complimented on her English skills, the success of which she attributes to her regular VIPKid sessions.

"She has the courage to talk to others and the interest to learn English," said Rina.

Firemen

The VIPKid model works because it is personalized, engaging, and effective. But no matter how successful the model is, technical difficulties will pop up for any virtual company. When relying on both the teacher and the student's computer capability and Internet access, connections do not always go smoothly.

The company actively works to create the best platform and technology to use when running its classes but also realizes that difficulties may happen. By knowing this, VIPKid has help on hand to troubleshoot these potential issues. These technology helpers, called "Firemen," are available to teachers and students during classes to help with any technology needs or issues.

Having support readily available takes the pressure off of teachers. Instead of stressing about possible tech issues, teachers are able to focus on what they do best: teaching. The addition of Firemen has complemented VIPKid's impressive commitment to offering technological support to teachers if they request it. Parents and students also appreciate having someone on hand to help with facilitating a virtual learning experience unlike any they may have participated in before.

Teacher Rebecca Phelps has appreciated the positivity that the Firemen bring to her experience on the platform. She remembered an experience teaching a rural open class and how helpful it was to have the support of VIPKid staff. At the start of that class, Phelps opened her screen and began teaching right away. After ten minutes of instruction, Phelps got an email saying, "You can start teaching now." She didn't know what to think, having already been delivering the lesson for ten minutes.

It turns out the wrong link was emailed out to Phelps's class, so the connection wasn't there and the class did not happen. Instead of getting flustered, the VIPKid staff and the tech teams were working furiously but calmly to solve the problem. After the chaos of the moment, Phelps sat down and read the messages that went on between this team of people when troubleshooting the problem.

"Every single one of them was encouraging," Phelps said. "There was never anything negative said. And to me, that's what it's about: bringing the positive to the situation. I left that situation and even though I was sad I didn't get to teach, I still left with a smile on my face. I knew that we gave it our all and we tried and we were going to do better next time."

CHAPTER 4
LEARNING WITH VIPKID

The VIPKid model enables students to learn English in a supportive environment. Students are supported through engagement strategies, building relationships with their teachers and participating regularly in short English lessons. All of these components help VIPKid lessons to be successful and set students up for learning. One of the most important pieces of this puzzle is VIPKid's curriculum and instructional approach.

Through VIPKid classes, students learn English through a total immersion model. This means that every part of every lesson is taught solely in English, even for students who have no knowledge of the language. The research on this model shows that is an effective way for students to quickly and effectively learn a foreign language.

Traditional language lessons teach a foreign language in small chunks, surrounded by instruction in the child's home language. But with total immersion, the amount of time that students are exposed to the foreign language greatly increases. It is the model that is embraced when people travel abroad and learn quickly by being surrounded by the language.

When students begin learning language with VIPKid, the lessons are twenty-five minutes and focus on developing speaking and listening skills. As students progress in their language skills, the classes become longer. Students in higher levels spend time both fine-tuning their speaking and listening skills and learning reading and writing skills in English.

The main course offering at VIPKid is called the major course line. When students first come to VIPKid, the intake process usually consists of trial classes and placement tests. Through the placement test and/or trial class, students' English proficiency will be assessed so that they will be placed at the appropriate levels. While many students start at the beginning of the major course line, some students have previous experience learning English and begin classes at more advanced levels.

Students begin learning in the major course line, which aims to improve the creative thinking of children ages four to fifteen. The major course line is broken into learning cycles, units, and then levels. There are roughly six lessons to a learning cycle and two learning cycles to a unit, making approximately twelve lessons per unit and then twelve units to a level. Currently, the curriculum is mapped out for seven levels, from Level 1 to Level 7 Plus. Level 7 Plus uses the Escalate English textbook from Houghton Mifflin Harcourt to bring language skills to children. The VIPKid major course curriculum is designed to take students through the completion of elementary school.

In Level 1, children engage in their first experience learning English. The goal in this level is to get students to love language learning and build basic skills. In Level 2, children younger than six years old learn basic skills, develop confidence, and practice speaking in English. Level 3 builds on this knowledge by focusing on reading and expression. At this level, students also engage in interesting and in-depth discussion topics.

At Level 4, students study fiction and nonfiction, understand different topics with the help of a small set of pictures, and discuss the key points of what they read. Level 5 has students take a leap in their language ability, cultural knowledge, and thinking ability. At Level 6, students develop their ability to read independently and use English to express their subject knowledge. A goal of Level 6 is to prepare students for the next stage of academic study or academic English ability required by international schools.

The VIPKid Level 7 Plus course promotes reading and writing, enhances listening and speaking skills, and lays the foundation for children's academic success in English. The English level of children who complete the

Level 7 Plus course is equivalent to students in fourth grade at an American elementary school.

The major course line was built to take into account the needs of learners in China, Korea, and other countries. Children's instruction is tailored to their English proficiency and cognitive level. In addition to improving students' language skills, the major course line is designed to teach students a global perspective and the skills necessary to become a 21st-century citizen.

In addition to major courses, VIPKid also offers personalized minor courses, self-learning tools like flashcards and digital libraries, and free resources. There are courses on topics including spelling, pronunciation, and vocabulary. Students can learn English through songs, virtual field trips, and other resources. As they make their way through the major course line, students can take these supplemental courses at the recommendation of a parent or a VIPKid employee called a Learning Partner, who works to create a personalized plan for students who are working their way through the VIPKid curriculum.

Learning Partners

In order to facilitate a smooth customer experience, families are matched with Learning Partners. These Learning Partners act as a consistent guide during students' time with VIPKid, helping students and parents plan out when they want to take classes and which classes they want to take. This is a particularly important role to have when students may not always have consistent teachers. Learning Partners provide consistency and guidance throughout students' journeys with VIPKid.

Learning Partners are matched with the families during the intake process, and they help to guide students along their learning progressions. Learning Partners can help parents make sense of the reports they receive from their children's teachers and plan next steps for how to support their children both through VIPKid and at home. In addition to recommending classes from the major course line, Learning Partners also recommend additional supplementary and fun courses. They might recommend that a child take a course about a mission in outer space or suggest a course specifically about phonics for students who need extra practice in that area.

Many Chinese parents may not be very familiar with different stages of language development, so the Learning Partners communicate information about language development as well as specific suggestions. For example, when students have just begun to learn English, they often go through a silent period. During that time, the students are not willing to talk in English, but they are still absorbing and learning the language. A lot of parents are worried about this phase, so a Learning Partner will help a parent recognize this stage and will give parents suggestions on how to positively guide their students along as they are working through that silent period.

Later on in development, students often get into a stage where some of their ability to follow grammatical patterns isn't as consistent and they experiment with different patterns of speech. A Learning Partner might help parents understand that phase so that parents who aren't experts themselves can have expert guidance along the way. Parents report that having Learning Partners helps to structure their experience with VIPKid and helps them to feel more confident and informed with the decisions they are making for their children.

Learning Partners primarily communicate with parents using WeChat, a messaging program that is very popular in China. Learning Partners are also able to send messages to parents through the VIPKid platform and sometimes check in via phone. Just as students build strong relationships with teachers, many parents build strong relationships with Learning Partners.

"We don't overly restrict the format in which they communicate, because some Learning Partners become really trusted confidants," said Stephenie Lee, product leader at VIPKid.

The role of the Learning Partners goes beyond just helping parents and students. Because Learning Partners have such close interactions with VIPKid families, they are also able to provide helpful feedback to the company. The Learning Partners have their eyes on the ground and are committed to making sure all students have a good experience.

"We internally build so many feedback loops for them to give feedback to other departments to make sure the entire experience is continuous and smooth," said Lee.

Pedagogical Approach

The VIPKid curriculum is immersive, meaning that it is taught entirely in English. In order to engage students in learning the language, the lessons need to be appropriately complex and enjoyable for the students. If the lessons were too hard, students would become easily discouraged and not have the stamina for learning the new language. If the lessons were too easy, students would not make as much progress as possible. It is an art to design lessons that help move students through their zone of learning. The VIPKid curriculum team has done a lot of designing and testing to make sure that their lessons follow a clear progression that is rigorous and engaging.

The VIPKid curriculum is rooted in real-world contexts, which helps students apply the language they learn to their lives. In addition to learning English, students in VIPKid classes also learn additional academic content. Lessons may be rooted in social studies, science, language arts, or math. By learning content through another language, students become more engaged in learning concepts *and* building their language fluency. They also pick up academic language, which can sometimes be forgotten in traditional language classes. Sometimes people see learning language and learning content as two separate things, but VIPKid hopes to teach students content through a foreign language. That way, students can make gains academically and in their language development at the same time.

"When we think about the introduction of nouns and verbs, for example, we're very much trying to introduce it in a way that uses an immersive English language learning context," said Stephenie Lee, a product leader at VIPKid. "Students learn through repetitive and differentiated exposure so that they can acquire and accumulate that language independently."

The VIPKid curriculum begins with concepts that are familiar and meaningful to students, such as days of the week, articles of clothing, and basic technology. By starting with things that they already know, students can focus on language learning through familiar content. As students progress, they may work on less familiar content as they continue to learn English. This method helps students to develop the language necessary to understand complex topics while engaging in progressively difficult tasks.

"When we think about how we contextualize the learning that students are supposed to be doing, we want to make sure, especially in the first couple of levels, that they are exposed to contexts that are relevant and familiar to them," said Lee.

The VIPKid curriculum is rooted in Bloom's taxonomy, a set of hierarchical models that classifies learning objectives based on complexity. For example, a task such as *defining* is classified as a lower-order thinking skill, while *analyzing* and *evaluating* are higher-order thinking skills. VIPKid lessons and slides are tagged by what level of cognitive complexity students are expected to engage in. Making the levels apparent helps teachers to understand the intention behind the academic tasks and decide how to challenge their students to use the appropriate skills.

In addition, VIPKid lessons incorporate a variety of scaffolds and supports that teachers use with students. Over time, the support that the teachers give students is gradually lessened as students become more independent. To understand this model, think about teaching a child to ride a bicycle. We often start by having a child use training wheels and holding the back of the bike as the child rides. As they become steadier, we take our hands off of the bike. Finally, when riding alone has been mastered, we remove the support of the training wheels. This is similar to how supports are introduced and removed over time.

"Teachers will sometimes use supports such as Total Physical Response, which is something we recommend a lot, particularly in the silent phase, to ensure that student and teacher coregulate and that they are able to stay connected and engaged as language development is happening," said Lee.

The VIPKid curriculum is based on an educational approach called the gradual release of responsibility, where the cognitive load shifts from the teacher to the student. Using this model provides a scaffolded approach where students are given more supports at the beginning of a lesson, then those supports are gradually removed as students develop independence and fluency.

Traditional lessons may be more teacher-led, with the teacher asking questions and the students answering. VIPKid's model aims to engage students in more complex cognitive tasks by providing them with the scaffolds to gain skills with the goal of creating independent learners. By taking this approach, VIPKid helps to nurture students and encourage them to take control of their learning.

Lesson Structure

When teaching VIPKid lessons, teachers are given suggested slides and scripts to use. The slides show up on screen with the student, allowing teachers to share visuals. The teachers can draw or write on many of the slides, providing an extra layer of interaction. When participating in a VIPKid lesson, teachers and students can see each other on the screen, as well as the slides.

Having consistent slides for each lesson helps teachers to come into lessons prepared and allows multiple teachers to work with the same VIPKid student. If a student worked on Lesson 3 with another teacher, for example, the next teacher could pick up at Lesson 4 without losing any time figuring out the content of student's prior lessons.

For the main course line, lessons follow a standard progression. Because the lessons are only twenty-five minutes long, teachers move through each part of the lessons quickly and efficiently. Each VIPKid lesson in the main course line starts with a warm-up segment, which is particularly important for hooking students' interest in the lesson.

"I want to start my class by really pulling in the student and getting them engaged," said teacher Rebecca Phelps. "If I can get their attention at the beginning, then I can just hold it for the rest of the twenty-five minutes."

After engaging students in a warm-up, the lessons work through various knowledge points broken down into pieces that students will learn. For a typical course, there may be three to four knowledge points per lesson. For example, the knowledge points might be vocabulary, phonics, and patterns.

Finally, there is a closing segment of each lesson where students can wrap up and review what they learned, including a good-bye wrap-up slide. During the course of each lesson, there are also opportunities for students to earn rewards. These rewards help to provide students with an extra dose of motivation to stay on task and perform well through the lesson.

Julia also mentioned that VIPKid helps kids to realize their dreams. Kids have lots of creative ideas about products, and when the Teaching and Research Team wants to design a new product, they are willing to use those ideas.

For example, Julia once said, "Claudia and Chris even gave recommendations when they attended EDU Learning Center meetings. I brought them into the office in Beijing during last year's summer holiday to have a special meeting with our product managers and technology engineers. And they eventually used Claudia's idea about the avatar scheme and upgraded the student's Learning Center."

Total Physical Response

Following a total immersion model can be overwhelming for students, particularly young students who might just be getting used to the concept of school. In order to help students through this project, it is important for them to receive a variety of supports when learning English. These supports consist of the use of visuals, movement, facial expressions, and modeling. By engaging with an English speaker through a structured curriculum, students are able to make gains in their language learning while having fun doing so.

One way that teachers can engage students is by using Total Physical Response to get students moving during classes. Total Physical Response teaches language by matching physical responses to verbal inputs. For example, a teacher may put her hand to his or her ear when indicating that he or she is listening to what a student should say. Using Total Physical Response helps students to make mind-body connections and increases engagement by getting students moving. Instead of simply sitting behind a computer screen talking, students who use Total Physical Response are able to move their bodies and get a more comprehensive experience.

Teachers may use Total Physical Response throughout lessons to increase engagement and language learning. For example, when learning about body parts, a teacher might say, "I move my hands," while making the corresponding movement. Then the teacher might ask students to move their hands, again making the same motion. As students make the motion, they can say, "I move my hands." By correlating movements with language, students are able to synthesize their learning. Making mind-body connections helps to cement language learning within students' bodies and make them more comfortable using the language.

Teaching Style

Teachers on the VIPKid platform are widely known to be energetic. As with Total Physical Response, teachers can make their faces very expressive. This is another way to use body language to teach verbal language. By being expressive and enthusiastic, teachers help to engage students physically and visually.

"When you put a really young child in front of a screen and you try to get them to learn something, they actually won't learn anything," said Rothla.

Instead, students learn best when they feel like there

is engagement and interaction between them and the teacher. By encouraging teachers to be lively, students are more likely to feel like the screen is removed. Instead of being a passive viewer of a screen, students become active participants in lessons.

Teachers on the VIPKid platform use a lot of visuals to make connections with real things that kids can associate with. Keeping it interesting and interactive is particularly important when helping young kids stay focused, so teachers may go through many different visuals and props through the course of one lesson.

"Just because teachers are energetic, that does not mean that they are tricking kids into believing that they're learning something," said Rothla. "Their energy serves the purpose of really actively engaging with the student."

The hallmarks of teachers on the VIPKid platform are energy, Total Physical Response, and props. Using props can help students connect with the teacher beyond the slides of the lesson. Many teachers use puppets, posters, and other realia in their lessons. Seeing which props different teachers have can be motivating for students and gives them a chance to interact with a wide variety of supports.

Due to the popularity and importance of these props, VIPKid decided to launch an online store where teachers can buy a variety of props to use in their lessons, such as VIPKid white boards and puppets of Dino, the VIPKid mascot. Before the store was launched, teachers would buy puppets and posters at a wide variety of different stores.

The VIPKid store has been incredibly popular with teachers. At VIPKid's Journey conference in Chicago, they set up a physical version of the store, to which teachers flocked during every break.

VIPKid gear has been gaining quite a following among both teachers and students. The VIPKid store now gives students the opportunity to buy toys and apparel for themselves.

"When a teacher is in a classroom and has a Dino doll, students can get very jealous because they want Dino, too," said Kevyn Klein, Global Director of Community at VIPKid, explaining the value of students being able to buy VIPKid products. "And you know what? It helps with engagement. If a teacher has a Dino and the student has a Dino, we can have the Dinos play."

Assessment

The scope and sequence of VIPKid's curriculum is designed to funnel students through different levels of language learning, gradually removing supports and encouraging independence as time goes on. But in order to effectively move students through the correct lessons, their knowledge and skills need to be regularly assessed. Even when lesson design is solid, that doesn't mean that students will learn at the same progression. Assessment helps to tailor VIPKid instruction and classes for each student. It is key to providing a personalized experience.

When students come to VIPKid, their English level is assessed through a placement test and trial class. In the trial class, the teacher gauges students' speaking proficiency and, for certain levels, there is a reading test that assesses reading abilities and comprehension. These assessments are used to provide a baseline level of students' language development. These initial assessments help to place kids in a lesson progression, but they are just the beginning of the assessments that students will experience throughout their VIPKid journey.

Once students get started using VIPKid regularly, they have formative assessments along the way. Every learning cycle, there is an English-focused formative assessment. At the end of twelve lessons, there is a slightly longer formative assessment that checks students' development over the course of those twelve lessons. In combination, these assessments provide an understanding of students' grasp of what was taught in those twelve lessons, every twelve lessons.

Though assessment is a large part of students' progression through VIPKid, the company is always thinking about how to refine and extend their approach. It is testing a proficiency test that can be taken as students make their way through course work, to get a stronger sense of students' holistic proficiency and development.

"One of the things that we find out from students

who take standardized tests elsewhere is that there's a lot of incidental language that's being used in VIPKid classes that the students also acquire," Stephenie Lee, product leader at VIPKid, said. "Right now with the formative assessments, we're capturing part of students' learning and we're ensuring that our learning objectives are met. This is also true in many ESL or EFL classrooms, but that is oftentimes the extent of the learning because they're also using Chinese to supplement. But in our case, because we have teachers and they speak English the whole time, there's just a lot of incidental learning happening."

These new assessments will help to measure this incidental learning. As Lee said, "To us, the way that kids learn through VIPKid is very much not limited to explicitly what we want to teach. We're excited to have a way to quantify and understand that progress."

Similarly to the anecdotal data that VIPKid gets from kids who take standardized English tests, they also get information from partner organizations. Sometimes there are VIPKid students who take tests or are involved in competitions elsewhere. VIPKid gets feedback from parents and sometimes receives anonymized data from partner organizations, which often show that VIPKid students are performing ahead of the norm. Through these data, VIPKid can compare its students' performance to that of other students who have been taking English elsewhere. Among students who have taken

English for the same amount of time, VIPKid students perform slightly better.

"We'll use that data as a way to understand whether or not our curriculum and instruction is hitting the goals that we'd like it to hit, but also whether or not there are some additional effects that we may not be currently capturing," said Lee.

Standards

In order to teach both academic content and language, VIPKid takes different academic standards into account when developing curriculum. When the curriculum for the main content line was created, it was aligned to the Common Core State Standards that many states in the United States have adopted. The Common Core State Standards cover reading, writing, speaking, listening, and mathematics.

In future revisions of the curriculum, the focus was aligning the content and instruction with the WIDA framework. WIDA stands for "Word-class Instructional Design and Assessment" and is an organization focused on the educational success of English Language Learners. By incorporating WIDA, the curriculum was able to be strengthened to include more language supports at a variety of stages of development.

But it's not just American standards that the curriculum team has considered. It is important to align the content of VIPKid lessons to what students already know or will be learning soon. With this in mind, the VIPKid curriculum team also takes into account the Chinese National Standards to get a better sense of what students are learning in school. VIPKid also collects feedback through learning partners and parents about what types of additional coursework students might be interested in.

"One of the two big buckets tends to be really extra-curricular extension, so into science or into other subject areas," said Lee, explaining the types of additional coursework that is in highest demand. "Then the other big bucket is specific subareas of English, such as grammar or phonics."

Learning Library

In addition to participating in VIPKid classes, students have access to a learning library with songs, stories, self-learning tools, and other resources to help them learn English. Kevin Gainey, content production manager, has helped to create a lot of this content. He has recorded songs and videos of himself reading popular stories in English, some of which are read by students every night, making Gainey a part of bedtime routines across China.

Gainey has also hosted many virtual field trips for VIPKid students, where he takes videos of himself in different places around the world. On location, Gainey acts as a travel host for students, taking them around and showing them different parts of cultures that they may not have been exposed to before. So far, he has filmed virtual field trips in California, Guam, and Mauritius. In California, Gainey filmed in San Francisco, Los Angeles, and Sacramento.

"I've always wanted to travel for my work, so this has been a great opportunity for me," Gainey said.

Next up for Gainey's virtual field trips include trips to Brussels and Dubai. Parents and students can watch these videos to learn more about different places around the world while also practicing understanding English. These videos are another way for students to engage in language learning in a different way.

All of the resources in the Learning Library, including songs, videos, stories, flash cards, etc., can be accessed by students and families outside of their regular class time. This gives an extra opportunity for learning language throughout the week, even when VIPKid classes are not going on.

Open Classes

Always focused on designing a personalized and effective learning experience, VIPKid started by perfecting its one-on-one lessons. But then VIPKid started wondering about scaling its impact. What if students could have additional classes during the week? What if there was a way to leverage the power of one teacher? The company started to supplement its one-on-one classes with courses in a one-to-many format. These larger courses are VIPKid's MOOC, or Massively Open Online Course, option. These courses are broadcast live, bringing one teacher in front of many students. Every month, at least a hundred open classes are broadcast to VIPKid students.

"Some MOOCs look and feel like a typical online course, with one teacher and a deck of slides," said Lee. "Some of them are like TV shows, edutainment basically, where knowledge points are embedded into story lines and then interactive questions are built in."

As VIPKid students take courses within the major course line, they can supplement their learning by joining open classes. These classes are offered at a variety of language development levels so that all students get a chance to engage in their content. Open classes aim to teach engaging topics, sometimes focusing on particular elements of American culture, such as festivals and celebrations.

VIPKid's MOOCs are also called "open classes" because students are welcome to sit in and join whenever they are available. These classes can have over a thousand students at a time, none of whom the teacher is able to see. This can be both invigorating and challenging for teachers.

"I really kind of just envision these ministudents staring back at me right behind my computer screen," said Rebecca Phelps, a teacher. "I try to be as animated as possible, as cheerful and engaging. You're not sure if the students are understanding or enjoying the lesson. But it's really fun, it's been a great experience to know that such a large number of students can have the opportunity to be exposed to VIPKid lessons." VIPKid has been developing many new online classroom functions for open class, such as collecting students' feedback data, emojis, AR functions, voice recognition, and interactions with the teacher. These new functions are not only helping in delivering positive user experiences, but also helping teachers to adapt classes more effectively according to the students' feedback during the lesson.

Junior Creator's League

As VIPKid has expanded, so has its course offerings. Stephenie Lee is currently working as a product leader for a curriculum line called Junior Creator's League. These classes are focused on STEM (science, technology, engineering, and math) and take a project-based learning approach.

Students in Junior Creator's League must have an English development level of at least four, which is equivalent to a first-grade English level. By having a basic understanding of English, students are able to better understand the subject matter and engage in the projects.

These classes take a constructivist approach, inviting students to experiment and construct their own ideas with the guidance of a supportive teachers. Through the Junior Creator's League classes, students are challenged to do things such as make scientific hypotheses and learn skills like coding. Engaging students in these projects provides them with a more personalized approach, as they come up with their own solutions and test out various ideas.

"I think she really hit the nail on the head with interest-based approach, and the way [Lee] set it up is also really cool and gives students a lot of autonomy and independence," said Madli Rothla.

The Junior Creator's League is just a taste of what's to come for VIPKid. They hope to develop more project-based classes in a variety of subject areas, knowing the power that this type of format has on kids' learning.

CHAPTER 5
RURAL EDUCATION PROJECT

The open class format is powerful because it can reach so many students at once. VIPKid has done a lot of thinking about how to bring its model to more students, particularly those who are impoverished or lack access to high-quality educational experiences. In 2017, VIPKid started the Rural Education Project. This project brings one-to-many classes to students in rural China for free.

The Rural Education Project aims to provide children in poverty access to high-quality English lessons. Through the program, rural schools in China are paired with a teacher who teaches lessons virtually to the class. Though it is a one-to-many model, classes in the Rural Education Project differ from open classes because teachers are able to see and interact with the students in their class.

The rural schools, which often have limited Internet and technology, are given the resources to make the online classes run smoothly. The Rural Education Project is the first step toward widening access to the successful VIPKid program that users know and love.

"We care so much about empowering *all* kids for the future," said Cindy Mi. "With technology, we see the possibility of working with more children. There are so many more kids we need to work with and help."

Seeing the resource disparity in rural and urban areas, Mi was motivated to bring high-quality English education to students in rural China. Many rural schools are short on qualified English teachers and resources for English learning. English is an important skill for all Chinese children, as it is not only a core course for Chinese students in grades three through twelve, but also is tested during the national college entrance exam and a vital skill to land a good job in the market.

Mi was further motivated upon hearing that once these students get to high school, they generally performed well. The issue rural students face is one of access; they do not have the same opportunities as other students in China. Knowing they have the resources and abilities to help students in rural China, Mi and VIPKid hope to change that.

Students in the Rural Education Project look forward to their classes with teachers. Amy, a third-grader from Yinhan Central Primary School in the Yaohui Township, Gaoling District, Xi'an, Shaanxi Province, is one of these students. At the end of 2018, Amy's class began participating in the Rural Education Project. After the first lesson, "How are you today?" Amy was inspired by her teacher Kelly's energy and enthusiasm. The lessons are fun and often quite funny, further engaging the students. The kids in Amy's class were so inspired to learn English that they even set up the Little Star English Community, where they come together every Friday to sing English songs, perform English dramas, and watch English cartoons.

Another student in the Rural Education Project, Wang Qianjin, is a fourth-grader in Longxi County, Dingxi City, Gansu Province, an infamous, poverty-stricken county with a local per capita GDP less than one tenth of Beijing. He has grown close to his English teacher, Ann, though they have never met.

"Teacher, you have brought me happiness, and you have a lot of knowledge. I am very touched when you give me classes. I like it very much," wrote Wang in a letter to his teacher. "Thank you teacher, thank you for teaching me knowledge. I understand the importance of these. I don't know how to show my gratitude, only to thank you on paper."

Just as the students love their classes, many teachers find working with the Rural Education Project to be the most rewarding part of their experience with VIPKid. Through their work with rural schools, these teachers are changing the face of education and providing English for these schools and villages that may not have otherwise had the opportunity to have such instruction. A bonus to working in the Rural Education Project is that teaching rural schools can happen in the evening, so the teachers can have one-on-one classes in the morning and then rural classes in the evening.

In 2017, teacher Kim Fortner began teaching classes as part of the Rural Education Project.

"It was just incredible," Fortner said. "You're given this class and a PowerPoint. Then the camera opens and you've got a room full of kiddos. Here you go! You think about classroom management and classroom engagement and you think, 'Okay, how am I going to do this? I'm a tiny little square.'"

But Fortner was surprised to find that she didn't need to use many of the classroom management strategies that she had perfected as a brick-and-mortar classroom teacher. The students in the rural class came ready to learn and very excited for their VIPKid classes. For Rural Education Project classes, the students' regular classroom teacher is also present during the lesson, but there is often a language barrier. Despite that barrier, Fortner has always felt connected and supported by the local teachers in her rural classes. Though they might not have the ability to collaborate with one another, Fortner sees the classroom teachers as partners.

"The teachers still have access to our recording and the material, so they are able to review the material if they want," said Fortner. "So we're supporting one another, but it's just across the world. It's incredible and unique at the same time."

The collaboration between teachers contributes to making the program a successful experience for everyone. One teacher whom Fortner knows had her students wear construction-paper hats with their English names written on them. These served as nametags for Fortner, who was able to easily see and learn the names of all the children in the class. Though Fortner is across the globe, she feels closer to the children she teaches when she can know their names.

Teachers have found many ways to make this long-distance collaboration work. For example, some teachers give groups of students names, like "The Tigers," to help facilitate group work with a virtual teacher. The teacher can then ask different groups to participate during the lessons by calling their group names. The extra effort that classroom teachers put into facilitating online lessons is very much appreciated by the teachers who teach the Rural Education Project classes.

Kim Fortner even got the unique opportunity to meet the teacher of one of her rural classes in person. In March of 2018, Fortner had the chance to travel to China. When she told the teacher of her rural class that she was coming, the teacher was very excited and decided to travel to Beijing to meet her. Fortner was overjoyed to meet someone in person with whom she had collaborated and grown so close on the screen.

During their meeting, Fortner shared gifts for the children in her rural class and asked the teacher to bring the gifts back to her class. Fortner brought each child a postcard with pictures of Fortner's family, pets, and home, and different coins, pencils, and pins. When the teacher passed out the presents to the students, they were very thankful and excited. The teacher even sent Fortner videos of the students opening the presents, which was invaluable to Fortner.

For many teachers, working with rural classes brings an opportunity to build more long-lasting bonds with students. They also feel thankful that their teaching is helping students who have fewer opportunities than others. Working in the Rural Education Project truly helps teachers feel like they are making a difference in the world.

Rebecca Phelps is another teacher who has benefited enormously from participating in the Rural Education Project. Like Fortner, Phelps began teaching on the platform as a one-on-one teacher.

"Never in my life have I been so nervous to teach before," Phelps said, remembering her first rural class. "I was literally sweating, holding all my props thinking I was going to drop them. But when I opened the camera and saw all those kids' smiles, it was like everything just went away. I hate to be biased, but the Rural Education Project is absolutely my favorite, definitely."

Phelps has found her time with the rural classes to be

nothing short of inspirational. She has been particularly motivated by the students' eagerness to learn and the commitment of the Chinese lead teachers. Teaching students who are so excited to learn has been a dream for Phelps, who currently teaches in five different rural schools.

The size of the classes in the Rural Education Project varies widely. For Phelps, her smallest class only has eleven students, while her largest class has over fifty students. The strategies Phelps uses with each class also varies, just like they would in a brick-and-mortar classroom.

"Even if you're teaching the exact same lesson, you have to be flexible," said Phelps. "You have to consider the classroom environment, the atmosphere, and how students respond. It definitely keeps you on your toes."

VIPKid has worked with numerous charitable foundations and NGO partners in a joint effort to reach more schools and improve the education quality for rural schools. While VIPKid funds the program, more investments in the Rural Education Project mean that more rural students can get access to VIPKid. The teachers themselves are so passionate about the Rural Education Project that they've even contributed their own money toward funding classrooms to take part in the project. At the Chicago VIPKid Journey conference, there was a silent auction to support the Rural Education Project. The teachers raised $1,500 (RMB10000) to support bringing VIPKid to rural schools in China, feeling thankful to have a way to give back to this meaningful program.

Meeting Students

Kim Fortner had the lucky opportunity to meet a teacher in the Rural Education Project, but Rebecca Phelps got even luckier: she got to meet her students. In early 2019, Phelps had a chance to visit one of the schools she teaches at for the Rural Education Project. As a result of teaching the same students for four semesters, Phelps has been able to develop strong relationships with the teachers and students in her class. She's been following the same class as they got older, and 2019 will be their last year at their school.

From the beginning, Phelps formed a really close bond with this class. She wasn't able to put her finger on what it was about them but has constantly been inspired by the students' excitement for learning. She also formed a strong relationship with the teacher, who—to Phelps's surprise—ended up being even more than a teacher: she was the principal, teacher, and the school cook. She even acted as the nanny for the seventy children at the school, as they live at the school during the week.

"She is the everything for the entire school," Phelps explained with admiration.

After such a successful experience, Phelps dreamed about visiting the school one day and meeting the students and teacher in person. On a whim, one day she sent the teacher a WeChat message and asked what she'd think about her coming to visit the school. The teacher was thrilled by the idea, which got her wheels turning even more.

Next, Phelps broached the subject of a potential trip with her Fireman, who had been helping her with that class since the beginning. The Fireman was also excited about the possibility, so Phelps moved on to explore the idea with VIPKid. To her surprise, several staff members offered to join her on the trip. Phelps and her husband were sold on the idea and began to make plans for their upcoming visit.

Because the school is in such a remote part of China, there were some logistics to work out. Phelps had to figure out where she would fly into and how she would get to the school. But, with some careful planning, they figured everything out, and in early 2019, Phelps and her husband set off on their adventure. They began by flying into Beijing and visiting VIPKid headquarters, where they were warmly welcomed.

"It was absolutely incredible to have several people from VIPKid along with us for the trip," said Phelps. "It wasn't expected at all, and to see the staff at VIPKid that went on the trip, I tear up when I think about it. They were so excited. Their passion showed through every single one of them."

After Beijing, Phelps, her husband, and the VIPKid team set off on their journey. First, they flew to Nanchong. From there, they got on a small bus and started on a journey into the mountains. Phelps remembers being unsure of how long the trip would take and how exactly they would get there. As they began to go up a mountain, the bus passed large plots of farming lands. Eventually, the roads got more and more narrow and precarious. At one point, the roads became too narrow for the bus to fit through. The entire team had to vacate the vehicle and walked the rest of the way to the school. Though it was nerve-wracking, Phelps remarked that the walk set the atmosphere the team was about to experience.

The team walked about two miles to the school, passing houses and people working in the fields. Every once in a while, a little motor scooter would pass them. Eventually, they approached the school. Seventy kids with matching jackets were lined up on the stairs, waving their Chinese flags and chanting, "Hello Teacher! Hello Teacher!"

"The principal of the school was there and comes up to me," said Phelps. "We just hug and smile. It was the most amazing thing I've ever experienced."

Phelps immediately felt welcomed by the teacher she had collaborated with during virtual classes. The school Phelps had interacted with so many times had made its way out of the computer screen and into real life. The relationships that Phelps had built with the teacher and students at the school were just as strong in person.

"What a surreal experience to finally be face-to-face with the adorable children that I had built a weekly relationship with for so long," Phelps wrote on her blog.

As Phelps walked around the school, she took it all in. The rural school was very different from schools in America. There was no heat, no air conditioning, and no laptops or other classroom technology. But Phelps was impressed by the color and general happiness that permeated the school walls.

Phelps and the VIPKid team spent the whole day at the school, getting to know one another even more. Phelps taught a mini-English class for the students at the school, which was a fun change of pace to have the English teacher deliver lessons in person. The twenty-one students from Phelps's sixth-grade class got together to sing songs, read books, and practice vocabulary.

The trip also consisted of some thoughtful gift giving. The VIPKid team delivered school supplies and food to the school. Phelps and her husband had gone to Walmart the night before the trip and bought outside toys, balls, hula hoops, and Frisbees to share with the school. Phelps wanted to give the students materials that they could play with outside. These supplies were all really appreciated by the teachers and students at the modest school.

The most touching part of the day was when the students from Phelps's sixth-grade class each came up to her and gave her something from their home. The gifts consisted of everything from a whole chicken to a bag of garlic to beautiful homemade flower arrangements. Phelps was incredibly appreciative of the thoughtfulness and generosity of her students.

"It was an incredible day," Phelps said. "I shared with them, they shared with me."

The students ate lunch with Phelps and the VIPKid team and did a performance for them. At this school, the students live there Monday through Friday. Since Phelps's visit was on a Friday, at the end of the afternoon, the grandparents who care for the children were walking up. Phelps remembers vividly the moment of having elderly Chinese women walking up with huge smiles plastered across their faces.

"We literally had to walk there, which of course the grandmothers did, too," said Phelps. "And here they are, in the midst of what we would consider hardship, and they had these huge smiles on their faces. That's exactly what I can say for the children at that school. They were so happy, so content, and so eager to learn. It was amazing."

The students at this rural school continue to be incredibly engaged with Phelps's classes. Recently, a VIPKid staff member sent Phelps a picture of the classroom as she virtually delivered a lesson. In the picture, you can see the classroom door. At the door, you can see that there is a collection of adults and more children peering in because they're wanting to learn English, too. The demand for learning English is so strong in the school, making Phelps's job even more rewarding.

"You're reaching out to so much more than what you think," said Phelps. "It's really mind-blowing."

CHAPTER 6
THE TEACHER EXPERIENCE

Many teachers come to VIPKid seeking out alternatives to classroom teaching. Maybe they want to make extra income outside of the classroom, maybe they want to spend time at home with their families, or maybe they want to be able to travel and live a more flexible life. Oftentimes, teaching on the platform allows these teachers to pursue a lifestyle that they formerly thought wasn't available to them.

The number of teachers on the platform who have been teaching for more than five years is as high as 70%, and the average length of teaching is seven-and-a-half years. Twenty-five percent of teachers on the platform have been teaching for ten to twenty years, and 7% have been teaching for more than twenty years. To use the platform, teachers are required to have a bachelor's degree or higher; however, 30% of teachers on the platform have master's degrees and 2% have doctorate degrees.

The top five majors among teachers on the platform are education, English, history, sociology, and music. The majority of teachers appear to live in Texas, Florida, Georgia, and North Carolina.

Teacher Kim Fortner was intrigued with the idea of being location-independent. In 2015, Fortner found herself temporarily relocating from Chicago to Florida for less than a year. As a classroom teacher, she would have had to get a temporary out-of-state teaching license in order to teach in Florida. Since she was not staying in the state permanently, she wasn't sure if it would be worth the time and effort. To top it off, it was also the middle of the school year when she moved, making her chances of finding a classroom teaching job more difficult.

Determined to stay in education, Fortner decided to look for options outside of the classroom where she could still use her teaching experience. She has her master's in education and was committed to finding something where she could still make a difference in the lives of kids. In researching flexible options, Fortner came across VIPKid.

"It sounded just like a dream," Fortner said, remembering when she first heard about VIPKid. "It was almost too good to be true."

Fortner signed up for VIPKid and has since spent three years teaching on the platform, which she calls "incredible." Eventually, she moved back to Chicago and continues to teach on the platform. Now, a typical day for Fortner involves teaching students one-on-one for about six hours each morning. She then teaches one-to-many open classes at night. In addition, she mentors several other teachers.

By teaching on the VIPKid platform, teachers are able to focus on what they really love: making a difference with kids.

"I feel like teaching with VIPKid has given me back the joy of what I loved about teaching and removed some of the stressors that you can experience at certain brick-and-mortar schools," said Fortner. "It has taken away different types of pressure, such as testing and paperwork. Because they are removed, we can focus on teaching and developing supports, whether it's a wonderful prop or somehow helping your student make connections. It's not on making copies or cleaning up the classroom."

Rebecca Phelps had a similar experience. She is the mother of seven children, with the youngest four kids adopted from China. Before teaching on the VIPKid platform, Phelps had over twenty-five years of experience in working with children, along with many hours of volunteering to teach English as a second language to both adults and children. When she started teaching on the platform, Phelps was looking for a way to stay home with her kids and raise money for an adoption. She thought VIPKid would be a way to earn supplemental income to put toward the adoption fund but never considered it would be something she would choose to do long term.

The youngest four of Phelps's kids have differing disabilities, which require frequent doctor therapy visits. In order to be there for her kids and support their needs, Phelps was thankful to find a way to work while still being present and actively involved in her kids' lives. She is able to teach in the morning before her kids wake up, then close her computer and be finished with work for the day—all before her kids have breakfast! In fact, for Phelps, teaching with VIPKid has become more than just a side job. It's what she considers "me time."

"My friends often ask me, 'You're so busy, you have therapy visits, you homeschool—when do you have *your* time?'" Phelps said. "It's in the early morning, I promise. Those couple of hours that I teach every morning, that time is for me. I literally wake up with a smile on my face, I'm so excited about it. The income is great, don't get me wrong. But to be able to give back and just share that teaching, it's been phenomenal."

VIPKid starts as a way for teachers to earn extra money, but it often becomes so much more. Teachers become members of the VIPKid community, feeling motivated and included in the success of the company. Many teachers begin by teaching a few one-on-one classes, but as time goes on, they decide to teach more classes and refer other teachers to the platform. This is vastly different from what classroom teachers experience.

For Rebecca Phelps, growing her teaching business took time, but it was well worth it. At first, she went full force into teaching students. She opened up slots, nervously started teaching, and soon her schedule was full. It has been full ever since, and she's loved every minute. Phelps has now taught the youngest students in VIPKid all the way through Level 6. She has taught students one-on-one, through open classes, and through the Rural Education Project.

Now, Phelps mainly chooses to teach older students because of the strong relationships she has been able to form with them. Phelps has been able to follow some of these students for almost the full two years that she's been teaching on the platform. She enjoys having a bond with her regular students and watch them grow in their English-language learning. It also benefits the students to have the consistency and support of one effective teacher.

"From the beginning, I absolutely fell in love with the kids. I love teaching, I love kids, I love teaching English. It's just my favorite," Phelps said. "I love the opportunity that VIPKid has given me to grow my business, and I love the opportunity to make the connection with people across the world."

Providing Feedback to VIPKid

Teachers are valued in all areas of VIPKid. The appreciation and inclusion of teachers is vital to the company's success. One way that teachers may play an active role at VIPKid is by giving voluntary feedback to the company.

Though teachers are provided particular curriculum to teach, they have opportunities to give feedback on the lessons and help to make them better.

"We try to understand how teachers are feeling about the content and give them mechanisms for voluntarily sharing their thoughts and experiences. This feedback allows VIPKid to improve," said Stephanie Lee, product leader at VIPKid.

Teachers are perhaps the best people to provide feedback on VIPKid products because they are on the front lines of using them. They have the background and experience to know what is working and not working, which can be invaluable to VIPKid. Even the best products can look successful in theory but have complications in the classroom that could not have been foreseen. By enlisting teachers as both the implementers and testers of the product, VIPKid gets guidance on how to consistently make its products the best they can be.

There are a variety of ways that teachers voluntarily give their feedback to VIPKid. In the very active online VIPKid community, teachers often talk about the curriculum, lessons, and platform. VIPKid staff members help collect suggestions as observers in those communities to gauge teachers' satisfaction and feedback. By doing this, VIPKid is able to get candid information about how teachers are reacting to its products. VIPKid is able to assess what is working and what are the areas for improvement.

In addition to talking with other teachers in the online community, teachers can submit feedback on particular lessons. At the end of every lesson they teach, teachers are able to submit feedback on the curriculum. They can highlight particular items that they would change or add to slides, comment on the flow of the lesson, or call out parts of the lesson that could be changed. This process helps the VIPKid team to gather important information that allows them to create the best teaching materials. But it also helps teachers to feel empowered and connected to the material they teach. While the curriculum is provided to them, the teachers are invited to have a say in how it is executed and improved, and they have complete discretion over how to teach the curriculum.

Another way that feedback is conveyed is through online tickets that go directly to the VIPKid team. This is a way for teachers to provide specific or aspirational feedback directly to the company. Within tickets, teachers can share their particular feedback, thoughts, or concerns. For example, teachers might notice a minor edit that needs to be made to a slide or an image that could be improved. But not all the feedback is constructive; sometimes teachers simply want to share ideas that they have been dreaming about. These can take the form of suggestions about curriculum, such as the desire to create a Harry Potter-inspired course about the science of magic. Other times, teachers submit tickets about technological functionality that they'd like to see or additional slides that they'd like to see created. Whatever teachers are hoping, dreaming, or wondering about, they are able to submit tickets to share their thoughts with the company.

Throughout the various ways that teachers can voluntarily give feedback, they are always encouraged to have open lines of communication with the company. If something is working well, VIPKid wants to know. If something needs to be fixed, they also want to know. By inviting all sorts of feedback, teachers feel that they are collaborative partners with VIPKid.

Creating Content

There are various ways to create content for students. One way is by creating user-generated videos that can be used by VIPKid students. This first started with virtual field trips, where teachers are able to share what their lives are like and give students more information about cultures and locations around the world. By opening up the creation of these virtual field trips to teachers, the company was able to widen their library and show Chinese students what teachers' lives look like.

For teachers, creating these videos can be like nothing they have ever done before. While used to teaching on video in an online classroom, many teachers have never created standalone videos on their own. But with the skills and confidence they have learned while growing their teaching businesses, many teachers are excited to give this type of video creation a try. They can watch other teachers' videos for inspiration and learn more about the types of things that they might include in their own virtual field trips. The collection of field trips has provided authentic content to students while giving them a chance to further connect with their teachers.

"We want teachers, many of whom have a lot of curricular experience, to share their skills and their life experiences with us," said Stephenie Lee.

Optional Resources for Teacher Support

Throughout their experience with VIPKid, teachers can receive guidance from others in the teacher community.

Almost every course line has mentors, who host optional workshops for teachers who are newly certified or who may want to learn more about a particular course line. Mentors share tips for how to teach the lessons in the course line well, including ideas about pacing or managing energy levels. They also share ideas about assessing or achieving different learning outcomes.

There are also optional workshops available for new teachers to take. The online workshops are a great way to become more acquainted with the different components of the VIPKid process. The workshops can center on a variety of topics, from extension in the online classroom to how to effectively use props. Currently, there are about sixty workshop topics that are available to teachers. Kim Fortner teaches several of these workshops, including classes on digital tools and Total Physical Response.

"We've got a new topic coming up where teachers are telling the community about Chinese culture," said Fortner. "We've got some teachers that are Chinese and are able to share with other teachers a little bit about Chinese culture so that they can bring that into their teaching."

Many of the workshop ideas come from teachers themselves. If there is a particular area that teachers are struggling with or a specific topic that teachers are wondering about, a workshop can be created to address those needs. Learning teacher-to-teacher is an effective and supportive model that brings out the best in teachers.

Teachers also have access to a comprehensive resource library, which they can refer back to any time at their discretion. The resource library is made up of materials and videos that teachers can engage with on their own time if they think it would be helpful. These resources center on a variety of different topics, from how to teach particular levels and classes to how to effectively implement Total Physical Response.

Introduction to VIPKid

With the introduction of Fast Pass, VIPKid thought about how to make a strong first impression with teachers. They have also considered how to welcome teachers into their community after they have signed up to teach on the platform. When teachers first signed their contracts with VIPKid, they used to only get one welcome email from the home office. Now, teachers receive a series of emails for them to read at their discretion, which introduce them to the company. The goal of the email sequence is to give teachers more information about VIPKid so that they can grow their business on the platform.

"A good example is that in February 2018 we filmed a video about the Teacher Service Team and we put it in a newsletter," said Kevyn Klein. "Anyone who came after February 2018 never saw that video. But now it's in the onboarding email series, and so every new teacher knows there's a Teacher Service Team and what they do. That consistency is really important."

The team also added checklists to each message in the onboarding sequence. This helps to encourage active engagement with the content. These onboarding emails have been very well received by teachers and have a high open rate. Klein attributes to this to teachers wanting to feel welcomed and encouraged.

Teacher Retention

The company has put a lot of thought into attracting high-quality teachers, but it is also thinking about how to get teachers to continue offering their services on the platform. In addition to attracting high-quality new teachers, VIPKid wants to be able to continue working with these teachers. Building a strong teacher community and valuing teacher feedback have helped with teacher retention. As teachers take on mentorship roles, they are able to use their experience in different ways. When teachers feel like they are part of a team and that their contributions are valued, they are more likely to continue growing their business on the platform.

Now, teachers take pride in how long they have been teaching on the VIPKid platform.

Six Apple Program

In 2018, VIPKid launched the Six Apple Program. This program brings teachers discounts to partner products and services, such as babysitting, technology products, Lingo Bus, and workout classes.

As part of the program, the team created a physical card for teachers to purchase. With these cards, teachers can get local discounts. In America, many retailers have teacher appreciation discounts that teachers can access with a card that shows where they teach. By selling teacher cards, this opened up a whole range of potential existing discounts for the teachers.

So far, the Six Apple cards have been incredibly successful. Ninety-nine percent of teachers who ordered a card have renewed the following year. Teachers appreciate the discounts, but they also appreciate the sense of community that comes with the card.

"I think there's something about feeling like they have joined the family. They have the card. They're part of it. Maybe it's not even the loyalty program, but it's a branding exercise. They're proud to hold that card," said Kevyn Klein.

Traveling to China

VIPKid sometimes gives teachers the opportunity to travel to China. Twice a year, VIPKid hosts a Beijing trip competition for its teachers. Twenty teachers are selected for an expenses-paid trip to Beijing each year. These teachers get a chance to visit the VIPKid home office and interact with members of the team while building community with other teachers. Often they get a chance to meet a few of their students in person, too.

But even without the Beijing Trip Competition, more and more teachers are traveling to China on their own. When they come to the country, these teachers want to be a part of the VIPKid experience. One of the things the VIPKid team recently introduced is a collaboration with a travel agency for teachers visiting China, including a one-day stop at VIPKid's headquarters. With this model, the teachers can come as groups once a month. By organizing the teachers into groups, they can have another opportunity to build connections with one another. When the teachers come to the office, they get a tour and a pack of gifts.

CHAPTER 7
THE TEACHER COMMUNITY

Though their teaching force is entirely virtual, VIPKid has put a lot of effort into ensuring that teachers feel supported and connected. This can be challenging, as teachers enter their own virtual classrooms from their individual homes. In the VIPKid model, there is no time built into regular classes for teacher interaction. As teacher Kimberly Purvis said, the biggest challenge of being a virtual teaching team is that there isn't a break room for teachers to hang out in and get to know one another.

Realizing this challenge of its model, VIPKid began to think about how it could help form a strong teacher community. VIPKid realized that the success of the company rests on the satisfaction and engagement of teachers. If teachers feel like a part of a community, they will likely open up more time slots and continue teaching on the platform.

"We were the first company to grow a teacher community," said Jessie Chen. "It helps teachers a lot to share with each other."

This desire to create a community amongst teachers led the company to form a whole team devoted to making sure teachers feel like they are connected to VIPKid. This team, called the Teacher Community Team, helps teachers build relationships with one another if they want to do so.

"For a lot of us, when we go to our jobs nine to five, there is this whole infrastructure built around the people that we meet," said Kevin Gainey, content production manager at VIPKid. "We talk to people physically, we interact with them, we share our goals. It's the whole water cooler thing."

This component naturally goes missing with those who are working virtually. After a twenty-five minute lesson, that's it. The teachers don't have a chance to follow up and create connections to the work they are doing. The Community Engagement Team represents teacher voices in product and policy decisions. VIPKid values the knowledge and experience that teachers bring and wants them to help inform its products. VIPKid is also constantly thinking about how its products and features add value to teachers.

The Community Engagement Team also supports the online community and helps to increase positive feelings among teachers. When a teacher feels connected, their satisfaction and effectiveness increases. They become friends with other teachers and feel like they are part of a larger community of engaged teachers. The engagement team thinks about three different types of connections: teacher-to-teacher connection, teacher-to-student connection, and teacher-to-company connection.

For example, teachers are able to make teacher-to-teacher connections through in-person meetups, where they get to know fellow teachers. Teacher-to-student connections are through ecards, where teachers and students can send each other notes outside of their regular class times. Finally, teacher-to-company connections are through optional events such as the Journey conference.

The Community Engagement Team also helps to facilitate ways for teachers to build their teaching businesses. The focus is on helping teachers grow their businesses and encouraging them to continue teaching on the platform.

Throughout the teacher journey, the focus is on personalization. All teachers have different personal goals, so the company thinks hard about how to personalize the VIPKid experience so that teachers feel successful. VIPKid wants its teachers to be active, engaged members of their teacher community, knowing the value this will bring to both teachers and the company.

"An engaged community is a multiplier to success. It can help you do more at scale, sign up less teachers, and be more authentic, and be more real," said Kevyn Klein.

How the Community Was Built

Though the teacher community is immensely strong, it wasn't always that way. When Klein came to the company in 2017, she was challenged to show proof of concept. Soon after she joined, in March of 2017, a Salt Lake City, Utah, teacher posted a comment on a newsletter inviting other teachers on the platform to come over for brunch. Klein knew that the three VIPKid cofounders would be in town for a conference at the same time as the meet up, so she invited them to go surprise the teachers. About thirty teachers were at the meetup when Klein showed up with the cofounders.

"We went to her house, and it was at that moment that I think we all realized the power of getting people together," said Klein. "They all shared their stories, and when they left that day, they went and referred more teachers. They went and connected more."

Jessie Chen remembers this moment well. Seeing the teachers getting together with one another changed the way that she looked at how the company could be supporting teacher communities.

"The teachers were really excited to see us. The three of us were very excited to meet teachers and understand their stories and user experience in VIPKid. We felt very connected," said Jessie Chen. "It's good for teachers to have an in-person meetup of their online community. We talked to Kevyn about going in this direction. We should help teachers build the teacher communities. This moment is how we started the teacher communities."

In February 2017, Mark Zuckerberg of Facebook announced a Community Summit. He formed the Summit out of the belief that people needed more connection to one another. A VIPKid teacher, Shannon, ran the largest VIPKid Facebook group at the time and applied to go to the Summit and was able to attend. To Klein, this cemented the idea that the teacher community was onto something. She began thinking about how the company could leverage its teachers' experience and enthusiasm.

In order to showcase teachers to the world, Klein decided to show more images of actual teachers in marketing materials and within VIPKid. When she arrived at the company, she saw pictures of the Beijing office and Beijing employees on VIPKid's social media accounts. There was one glaring omission: teachers. Klein began to make sure that pictures of teachers were taken at all meetups. These photos were used to show to potential candidate teachers, making it clear that there are other teachers like them teaching on the platform.

In 2018, the Advocacy and Care team was created. Through the work of this team, VIPKid was able to send hundreds of care cards to teachers when something momentous happened to them. These handwritten cards are sent to teachers during all times of life—in the happy moments, and sometimes in times of sadness. Whatever the occasion, the goal of the care notes was to make teachers feel more than just a number, but like they are a valued member of the VIPKid community.

The Journey conferences, a series of large in-person events put on by VIPKid, were also launched in 2018. Two hundred and fifty teachers attended the first Journey conference in Salt Lake City. Like many things at VIPKid, the idea for the conference came from a teacher. The teacher wanted to arrange a large gathering, either with or without the company.

"Well, if she's going to do it, we can also support this and make it bigger," Klein remembered thinking.

This teacher's idea and motivation pushed VIPKid to host the first Journey conference. They had never done anything like it before.

Journey Conferences

In 2018, VIPKid held three regional Journey conferences in Salt Lake City, Dallas, and Orlando. Teachers flocked together to make connections and celebrate their experiences with VIPKid. The conferences were like big professional development sessions with a healthy dose of team building worked in.

"We are able to make those in-person connections. It's been great," said teacher Kim Fortner. "Eventually the conferences are going to spread out because we're getting teachers all over the world."

The Journey conferences have been very popular

among teachers. So far, about two thousand teachers have attended a conference. The teachers love having the chance to connect, further their learning, and meet members of the VIPKid team.

"When we post about the conference, it sells out in two hours. It's amazing," said Klein.

It's not just teachers who come to the Journey conferences. Reporters and the VIPKid public relations team from China come, too, aiming to share the conference with VIPKid parents and students. VIPKid staff also enjoy coming to the conference and getting the opportunity to connect with the wide variety of enthusiastic teachers.

Each Journey conference begins with a keynote address from a thought leader. In the past, these keynotes have been delivered by Cindy Mi, former First Lady Mrs. Laura Bush, and former Chicago Mayor Rahm Emanuel. Each conference has a theme, and for the rest of the day, the teachers and staff lead workshops on that theme.

"Hands down, the teachers love meeting staff from VIPKid headquarters the most," said Klein. "When we survey them, that's what they say. They say, 'I love meeting staff. I love talking to them and feeling connected to them.'"

Klein's team has used a Net Promoter Score to assess the success of the Journey conference, and the response has been remarkably positive.

Even a few VIPKid students have gotten a chance to participate in Journey conferences. At the Journey conference in Orlando, eleven-year-old twins and VIPKid students Chris and Claudia were the masters of ceremonies. They introduced the mayor pro tem of Orlando and the events of the conference. The twins even copresented a session with their mother, VIPKid VP of User Experience Julia Lee. During their session, they talked to over two hundred teachers about the student and parent experience with VIPKid. After the presentation, several teachers were so charmed by the twins that they decided to become certified in Level 7 (the VIPKid level Claudia and Chris were currently on) so that they could have the opportunity to teach them.

Online Collaboration

Teachers may only have a chance to participate in a Journey conference occasionally, but they have access to regular online collaboration with other teachers. Teachers have a variety of ways to interact with one another online, from a VIPKid discussion board called the Hutong to groups on Skype and Facebook.

"It's fun to have the space so that we can get to know one another. When I first started teaching with VIPKid, it was helpful to learn from teachers who had been doing VIPKid for a while. It was so different going from a live classroom to an online classroom, and it kind of scared me," said teacher Rebecca Phelps. "Just to think through and to see what they've done that's worked and even hear from the teachers that were brave enough to share about the things that didn't work. That, to me, was just invaluable."

There are even smaller, more intimate communities for different subgroups of the VIPKid community. For example, there are Facebook and Skype groups for teachers in the Rural Education Project and teachers who teach open classes. Through these groups, teachers are able to get more specialized support from people who are going through the same experiences as they are.

Though VIPKid's first teacher Madli Rothla is no longer a teacher (she now works on the Academic Planning Team at headquarters), she sometimes sneaks into the teacher's forums to lurk when she needs a "feel-good" moment. In the forums, teachers share funny things that students say, struggles they're having, gifts their students sent them, and talk about basically anything related to VIPKid.

"It's so cool to read the teacher's stories sometimes," Rothla said. "There's a lot of positive feedback and a lot of positive stories. The teachers are really a community."

What can the change in positive sentiment be attributed to? Increasing positive feelings expressed in online comments can be a tricky pursuit. The biggest way to increase this number is by increasing teacher happiness across the board. When teachers feel supported and listened to, they're likely to show up on the Hutong sharing those positive experiences.

"It's really powerful because everyone understands

community," said Klein. "They understand what they do in the community, and when someone does something good, just like in a classroom with students, they reward them.

"It shows the power that on our own, we can't do it, but using the community, we're able to actually amplify and increase our positive sentiment."

Meetups

After crashing the Salt Lake City meetup in 2017, Jessie Chen has worked with Klein and the Teacher Community Team to increase ways for teachers to collaborate in person.

"Teachers are located anywhere in the world, and it's possible some teachers may feel isolated or alone," said Chen. "At in-person meetups, the teachers get to see their friends, get to know one another, and share their stories."

Each month, teachers get together in person to connect at regional meetups, which they plan on their own. A hundred regional meetups happen each month, bringing together teachers to share about their experiences with VIPKid.

"Our meetups are get-togethers where we do anything from have pizza to do a prop swap," said Kim Fortner, who helps run a monthly meetup in Chicago. "For the prop swap, you can bring props you've been using in your classroom to trade. So if you've been using the same puppy for two years, maybe you can find something new to use."

The teacher community works to make these meetups as helpful as possible. For example, in February, many Chinese families are traveling because of Chinese New Year. This means that teachers may teach fewer classes than they're used to during this holiday period. To keep up positive energy amongst the teacher community, many meetup organizers arranged their monthly events at Chinese restaurants. In February 2018, there were one hundred different events happening at Chinese restaurants. Teachers were making dumplings, enjoying food together, and learning about Chinese culture.

In Kansas City, a teacher meetup even brought in people to teach them more about Chinese culture. TEDx speaker Nick Wang shared his experience growing up in China, an assistant professor from TESL presented, and an associate dean of Asian Studies spoke, as well. All of these events were organized by the teachers and cost no extra money.

In addition to being a way to further their professional learning, these meetups are also an opportunity to be with teachers who truly "get" the VIPKid lifestyle. They are a chance to be together with people who understand things like Total Physical Response and are learning about Chinese culture at your side.

"We get to get out of our online world and out of our home offices and meet with one another in real life. We can talk all things VIPKid because eventually your family is kind of like, okay, I don't get it, this lingo you're using," said teacher Kim Fortner.

Supporting One Another

When teachers come to Kim Fortner's meetups, she always asks them how they heard about the company. Most often it is through word of mouth. Most teachers come to learn about the company through a referral, which then ensures that they have at least one connection when they join.

Fortner initially found the company through Google and, at first, thought it might be too good to be true. Over the past couple years, Fortner has seen teachers become more comfortable trusting the company.

"A few years ago, all kinds of people were just like, 'What is it that you're doing and who do you work for?'" Fortner said. "Now I feel you always meet someone that knows someone or a neighbor, or their dentist's son or daughter or something. There's now this connection, which is really great. We're growing so fast."

This support continues through teachers' time teaching on the platform. Teachers regularly interact through social media, including various Facebook groups and meetups. But another practical way teachers end up helping one another is by pitching in with substitute classes for one another. This happens for both one-on-one and one-to-many classes.

"If you're one of the teachers that can offer to substitute when something comes up, you can jump in and meet a new group of kiddos and experience a new school," said Fortner. "That's really great about what we have in the teacher community that we can support one another in that way and make sure that class has a teacher for that day."

One of Kim Fortner's main concerns when she joined VIPKid was that she wouldn't have enough opportunities to be social. But Fortner has been pleasantly surprised. In her four years teaching on the platform, she has made friends around the world, met up with other teachers, and built relationships with teachers both online and in person.

"I hesitate to even call it a job," says Fortner. "I'm able to connect with other teachers and make a difference with students."

CHAPTER 8
CULTURAL CONNECTIONS

Both teachers and students value the cultural connections they are able to make through VIPKid. Not only are kids able to learn English, they are able to make friends across the world. The cultural connections also help to support students' English development as they learn vocabulary and incidental language related to American culture.

The VIPKid curriculum builds in opportunities for both children and teachers to share about their culture. There are many opportunities for cultural sharing, such as through holiday and festival-themed curriculum. For example, in recent years many teachers got into sharing Thanksgiving with their students, even using real food, like pumpkin pie, as props during their lessons. For Halloween, teachers dressed up and even inspired some of their Chinese students to come to classes in costume. Through the curriculum, students and teachers have many opportunities to share and learn about each other's cultures.

VIPKid students Claudia and Chris have been particularly interested to learn about American holidays. They have valued the opportunity to learn about different winter holidays, such as Christmas, Hanukkah, and Kwanzaa, that Americans may celebrate. But it's not just learning about another culture; students also appreciate having the chance to share about their own culture.

"I got better at understanding my own culture and a better recognition of my own culture," said Claudia.

Like Claudia and Chris, teacher Rebecca Phelps's students have been very interested in learning about life in America. Typical questions that she gets asked involve Los Angeles and New York. Based in Portland, Oregon, Phelps isn't able to make personal connections to these places, but she is able to share more about her hometown. She loves to show the topography around where she lives, like waterfalls, mountains, and snow.

When you are surrounded by your own culture and home environment, it is hard to realize its defining characteristics. Interacting with people from different cultures gives students a chance to get a stronger understanding of their own culture and environments as they compare and contrast their experiences with those of their teachers.

For teachers, making these cultural connections is also incredibly rewarding. For Kim Fortner, teaching with VIPKid makes the world seem much smaller. When she teaches students at 9 a.m. in China, then goes to have dinner in Chicago, she feels connected with the global community.

"Now I feel like I have families in China," said Fortner. "I've seen my student Amy's little sister grow up—she's two now. I know the family members' birthdays, and they know my birthday and my pets' names. I truly feel as though I have family across the world."

It is incredibly rewarding for teachers and students to feel these global connections. For Rebecca Phelps, this connection is particularly personal. Phelps has adopted four children from China, so being able to learn more about Chinese culture and build relationships with Chinese families has been invaluable. Phelps has learned about all sorts of ways that daily life is different for Chinese students, from what happens in traditional classrooms to where her students go on family trips. Learning more about life in China has helped Phelps connect with her own children's heritage. She is able to pass on what she learns to her children who were born in China to help them learn more about their native country.

"Being able to have this connection with China while being so far away has been really tremendous," said Phelps. "We love China, our family loves China, we love the people of China. Being able to connect with children and their families has been such a huge blessing to me,

Cindy Mi, Founder and CEO of VIPKid.

VIPKid's first class.

Early students took photos with VIPKid staff in the office.

For the Spring Festival of 2019, more than twenty Chinese students came to California to spend the New Year with teachers.

A boy and his teacher wrote the Spring Festival couplets together in Chinese and English.

At the 2018 Shanghai International Children's Book Fair, VIPKid announced a strategic cooperation with American Scholastic Inc., the publisher of the Harry Potter series. "Harry Potter's magic English class" became the highlight of the fair.

On China's variety show *The Sound*, a clip of the famous animated film *Frozen* was dubbed into English by two young students from VIPKid, which astounded the audience.

On March 10, 2018, the first Journey conference was held in Salt Lake City, at which nearly three hundred teachers read the Tang Dynasty poem, "Ascending the Stork Tower" aloud, wishing Chinese students to "go up a story still higher" at the start of a new semester.

The postcards for the Year of the Dog that Chinese junior students drew for their teachers were very popular at the Journey conference in Salt Lake City.

On August 8, 2018, a Journey conference was held at the George Bush Presidential Library & Museum in Dallas, Texas, with more than four hundred teachers present.

On March 9, 2019, six hundred North American teachers gathered in Chicago, Illinois. Chicago Mayor Rahm Emanuel presented at the teachers' conference and delivered a speech.

VIPKid Founder and CEO Cindy Mi together with teachers at Journey Chicago.

In 2016, all VIPKid staff attended the annual meeting in South Korea.

On August 23, 2017, VIPKid held a press conference for the "Opening Ceremony of the New Semester," in which Cindy Mi announced that the company had completed its D round of financing with new investment totaling 200 million US dollars.

VIPKid staff at the press conference for the D round of financing.

On August 2, 2018, at the press conference for VIPKid's D+ round of financing, Cindy Mi announced the "V+ Strategy" with education, technology, and service as its core.

At the 2018 Annual Gala, Lingo Bus Director Su Haifeng and his staff performed together.

At the 2018 Annual Gala, VIPKid's three cofounders (from left to right: Victor Zhang, Cindy Mi, and Jessie Chen) smiled when they were invited to the stage.

On October 18, 2018, the three cofounders presented at VIPKid's fifth anniversary.

The 2018 Annual Gala was held in Tianjin Wuqing Gymnasium, attended by employees from Beijing, Shanghai, Chengdu, and Dalian, with more than five thousand people on-site.

On January 5, 2019, VIPKid held its annual conference accommodating ten thousand people in Beijing's Wukesong Gymnasium.

Fatu Primary School in Yudu County, Ganzhou City, Jiangxi Province, was admitted to the VIPKid Rural Public Welfare Project in 2017. The children were looking forward to introducing themselves to their teachers.

On April 16, 2018, VIPKid volunteers went to Shujie Primary School in Dali, Yunnan Province, and held a poetry recital for children in the countryside.

Rural Education Project teacher Jessica surprises students with an in-person visit to Shujie Primary School.

For the first time, Jessica didn't give lessons online but taught the children in a classroom.

VIPKID is a leading brand which focuses on online English education for 4-12 years old Kids. The company offers 1-on-1 English language instruction provided by highly qualified teachers from North America. The course materials are aligned to U.S. Common Core State Standards(ccss), and with immersive English environment, one effective second language teaching method VIPKID helps Chinese children truly realize the dream of "learning American elementary school courses at home".

To date, VIPKID has more than 200 thousand paying students and 20 thousand teachers from North America and Canada. The renewal rate of students is higher than 95%, and the revenue target of 2017 is 5 billion.

On 23rd August,2017,VIPKID finished the D ound finan and receives $200 million. This e, the in ment was led by Sequoia Capital Tencen ollowed as a strategic investor. YF tal,M Partners China,Zhen Fund and o Ca also jointly participated in.The $200 ion raising was also the largest one in bal investment field. Lingo Bus, the first en Chinese education platform was cetld to officially operate on the

In December 2017, Lingo Bus made its debut at the 4th World Internet Conference held in Wuzhen, Zhejiang Province.

In April 2018, Cindy Mi attended the Boao Forum for Asia and outlined her vision for the future of educational technology at the New Deal Sub-Forum on Private Education.

In April 2018, GSV CEO Michael Moe put the VIPKid story in his keynote speech at the opening ceremony of ASU+GSV, the world's highest-level education technology summit, as a case study of Sino-US education cooperation.

and it's been a huge blessing to really get to know China better."

Phelps has been particularly impressed by the dedication of her Chinese students, sharing how she learned about the importance of the education system in China. Seeing how hard her students work in twenty-five minute classes is inspirational to Phelps. By knowing the effort her students put into learning, Phelps feels pushed to give teaching her all. Many of Phelps's students are studying many different things and putting effort into developing their abilities in diverse areas. For example, one of her six-year-old students is studying the piano and recently played Phelps a piece. Phelps was blown out of the water by the student's abilities; as the student played for her, Phelps sat on the other side of the screen with a huge grin on her face.

"I have a map in my classroom-slash-office and sometimes just literally sit here and look at the distance," said Phelps. "China doesn't feel that far away because of VIPKid, and that just means the world to me and to my family."

Cindy Mi, Founder and CEO of VIPKid, has never underestimated the power of student-teacher relationships, but even she was surprised by how meaningful and magical these virtual interactions have become.

"How strong the connections are has gone beyond my imagination," said Mi.

CHAPTER 9
LINGO BUS

VIPKid has found great success teaching English to Chinese children. This led to the question: could the same model be used to teach Chinese to English-speaking children? The company knew that its platform and total immersion approach to language learning worked, and Mi had always had the vision to build a worldwide, cloud-based classroom. Through the development of VIPKid, the company had created effective e-learning technologies that could be applied to new contexts. It decided to explore teaching Mandarin to English-speaking children with its Lingo Bus brand.

Lingo Bus launched in 2017 and by 2019 had more than twenty thousand students from 104 countries. It was the first company to offer total immersion Chinese classes online. Lingo Bus foresees a bright future for itself. Because of China's growth, more people will want to learn Chinese.

To research the idea of Lingo Bus, the company focused on North America, because there is a large number of teachers on the VIPKid platform in North America who have children. The company surveyed parents in North America and found that they were more enthusiastic about learning Chinese than expected. In the pilot phase, users from North America and countries such as Japan, Germany, and India were selected.

Lingo Bus works very similarly to VIPKid, teaching Mandarin to children ages five to twelve from the comfort of their own homes. When parents sign up for Lingo Bus, they are able to take a trial class to see what it is about. Then children are placed in leveled tiers of classes to support their Chinese language development. The Lingo Bus classes are fully immersive, taught entirely in Mandarin from the very first class on. The teachers use similar strategies to VIPKid, such as Total Physical Response, and bring incredible energy to engaging students in learning a foreign language.

Though the programs are very similar, there are several differences between VIPKid and Lingo Bus. One thing is that Lingo Bus has two main courses. The "Chinese as a Second Language Program" like VIPKid, teaching Mandarin to students whose families have no Chinese background. These students start to learn Chinese from scratch. The focus of the Chinese as a Second Language Program is teaching speaking and listening skills.

The other program, called the "Chinese Heritage Program," teaches Chinese and Chinese culture to foreign students of Chinese parents. Most of the students are from America and Canada but often speak some Chinese at home. Because they may already have some knowledge of Mandarin, these students focus more on reading and writing than students in the Chinese as a Second Language Program. Many parents sign their children up for the Chinese Heritage Program to help their kids connect with their Chinese culture, even when living abroad.

Lingo Bus is the first education company to use total immersion to implement a system of Chinese courses. Its approach is different from previous approaches to language learning, where there was no authoritative method for teaching children Chinese. Platforms like Lingo Bus are creating a standardized approach to the teaching of Chinese that didn't exist before.

Haifeng Su is the general manager of Lingo Bus. Before coming to the company, he was a Chinese teacher and had worked with international schools. One thing that he appreciates about Lingo Bus is the potential it has to reach so many students. While brick-and-mortar schools can only cater to a limited number of students, the audience for VIPKid and Lingo Bus is unlimited.

"This is exciting," said Su. "With high-quality content and curriculum, we should try to improve efficiency and help more children."

The Lingo Bus curriculum puts a big focus on engagement. When learning about Chinese characters, students have pictures that go along with the characters. The students learn the structure and meaning of characters together. Su said that, as a child, he learned Chinese characters by repeating characters, writing the same one over and over. But he found that hard to remember because he didn't understand the meaning of the characters or why they should be written as they are.

"With Lingo Bus, the students see the characters part by part," Su said, explaining the difference between the company's approach and the approach he experienced when in school. "The students see the parts and why they fit together. It makes them easier to remember."

Learning characters is a common frustration amongst those learning Chinese. One Lingo Bus teacher, Du Cong, worked with a four-year-old Chinese girl living in Europe. Her mother insisted that her daughter just learn listening and speaking skills, but that there was no need to learn Chinese characters. Cong persisted, convincing the mother and child that learning characters can be fun. She used various approaches, such as combining actions, pictures, and stories together to show the parts and meanings of the characters. The girl became motivated to learn characters by playing a Chinese character game during her Lingo Bus classes.

The Lingo Bus curriculum is informed by a variety of standards and frameworks. The Lingo Bus curriculum is designed to meet the National Standard for Language Learning and the 21st Century Skills for World Languages. The materials are aligned to the ACTFL (American Council on the Teaching of Foreign Language), the Youth Chinese Test, CEFR (Common European Framework of Reference), and HSK to ensure high-quality content in each level of its curriculum.

Teachers on the Lingo Bus platform share many of the same satisfactions as teachers on the VIPKid platform. Before coming to Lingo Bus, Liu had taught Chinese to adult learners. She remembers how much work it was to prepare and teach language classes in this setting. But Liu felt trapped by her job options. As Chinese language teacher, she only saw one way to use her experience: become a university teacher. When Lingo Bus began, Liu was excited to explore options outside of the traditional classroom.

"With the Internet, we can connect with the world," said Liu. "Lingo Bus is helping teachers have other ways to teach."

Lingo Bus helps teachers focus on what matters most: making an impact with their students. Instead of the traditional preparation and report writing that goes into teaching, Lingo Bus sets teachers up with high-quality curriculum and processes. Students are given standard preview and review materials to work with before and after lessons. The teachers write a short report to parents after each lesson, but otherwise teachers' jobs can be completed entirely within each twenty-five-minute Lingo Bus class.

Teachers are also given the guidance needed to successfully teach with Lingo Bus. To meet the needs of teachers, Lingo Bus offers regular workshops on topics like immersion and Total Physical Response. Through the teacher service platform and the resource center, Chinese teachers around the world are provided with teaching resources and support.

The demand for teaching at Lingo Bus is high. Su said that there are many excellent Chinese teachers, and many people are enthusiastic about the platform. Every day, a large number of teachers who are excited about teaching Chinese and Chinese culture are applying, but Lingo Bus has rigorous selection standards and only accepts the most qualified candidates. Most teachers are based in China, but a few are based in other countries.

The Lingo Bus team has been taking what has worked in the VIPKid community and applying similar approaches to their teaching community. Lingo Bus teachers are encouraged to collaborate. They connect via WeChat and DingTalk, talking about how to improve teaching methods and giving feedback on the Lingo Bus curriculum. The teachers are able to provide one another with support and encouragement along their Lingo Bus journey.

"We are a big family," said Su.

Recruiting students can be a bit harder, but Lingo Bus, like VIPKid, relies heavily on referrals. If students and their parents enjoy the program, they are rewarded for referring their friends. Lingo Bus also attends global conferences to help more people become aware of Lingo Bus and their model. They are also looking for more global partners to spread the word to even more students and families.

These cultural connections that are made on screen have a big impact in the real world. For example, Finley, a Lingo Bus student, had a Chinese student transfer to her school. The new student was an eleven-year-old girl from Zhengzhou who was just adopted a week before starting school. Excited to use her knowledge of Chinese, Finley used all the Chinese (like "above" and "below") she had learned in Lingo Bus classes and asked the new girl about the front and back of coins. The new girl smiled happily, excited to hear the familiar Chinese. Soon they became friends and played a coin game in the classroom. Because of Finley's knowledge of Mandarin, the Lingo Bus student was able to communicate with the new student and make her feel comfortable. They became great friends, with the American student going a long way to making the Chinese student feel more welcomed in her new community.

Lingo Bus students appreciate knowing how their language learning can be an asset in real life. They also love to make cultural connections and learn more about Chinese culture.

"Now I'm super excited about Chinese New Year. I

think it's good that we learn about their things and they learn about ours," said Elizabeth, a Lingo Bus student. "We should do more Chinese stuff, and other countries should do more American stuff. You know, maybe if all the countries cared about learning about one another, everyone would get along better."

Sometimes the cultural connections made through lessons are unexpected. When Du Cong, a Lingo Bus teacher, showed an eight-year-old boy a red envelope during a lesson on China's Spring Festival, it triggered the little boy's memories of his grandfather.

Before this, Cong did not know that the grandfather of the boy was Chinese. The red envelope reminded the boy of a time when his grandparents gave him an envelope.

"I was very touched at the time. A little red envelope awakened a memory of his family," said Cong. "I listened to his story and felt very warm. We live in different countries, but I know his feelings."

Lingo Bus students are also given opportunities to further their cultural knowledge and share with others. In the curriculum, Chinese traditions, literature, and architectural objects are chosen to be "cultural points" for each class. The goal of the curriculum is to emphasize cultural awareness.

One time, Lingo Bus hosted a video call for students to recite Chinese poetry. Among the participants was a young student whose family of nearly ten people stood in front of the camera and recited the poem "The Geese," by the early Tang poet Luo Binwang. These types of events create further opportunities for cultural connections.

Knowing the power of the platform, more than two thousand teachers who teach for VIPKid have signed their own children up for Lingo Bus. While these teachers are teaching Chinese children English, their own children are learning Chinese. Su is hopeful that this cultural exchange will create a future generation of global leaders and ambassadors.

Teachers on the VIPKid platform are encouraged to sign their children up for Lingo Bus classes by being offered a discount by VIPKid's Six Apple Program. Heidi Demaio is a teacher who has enrolled her son in Lingo Bus classes.

"I want my son to learn Mandarin in a one-on-one environment. I've seen it be such a special tool for the kids I teach for VIPKid that now I want the same for my own son," said Demaio. "One of the most special things about Lingo Bus is the teacher's connection with the students. The ability for them to hear pronunciation, but also use the media that goes along with the relationship. Those two things coincide really well."

Demaio's son looks forward to his Lingo Bus class, always asking when his next class is. A bonus of the Lingo Bus experience is that Demaio is able to learn alongside her son. By sitting with him during the Lingo Bus lessons and talking with him afterward, Demaio has been able to learn some Mandarin. This in turn will help her to make stronger connections with her VIPKid students.

In the future, Lingo Bus plans to increase enrollment and scale their offerings. Though they are focused on developed countries right now, Su hopes that they'll be able to create more ways for children from developing countries to learn Chinese. Perhaps one day there will be a branch of Lingo Bus similar to the Rural Education Project.

For now, Su has his hands full with increasing the reach of Lingo Bus. But he wouldn't have it any other way. When asked what he enjoys most about running Lingo Bus, Su summed up his feelings with three words: "passion, love, and fulfillment."

CHAPTER 10
THE FUTURE OF VIPKID

The VIPKid vision is to be the leader of global K-12 education. VIPKid is well on its way down that path but is always thinking about the future. Cofounder Jessie Chen is invested in finding ways to leverage VIPKid's cloud-based platform.

"Someday, no matter what children want to know, they can have VIPKid cloud-based classes available to them," Chen said.

Though the company is focused on English learning, there is talk about expanding to other subject areas, ages, and languages. Though the company is thinking about ways to expand, they are also constantly focused on improving what they have. Kevin Gainey, community production manager at VIPKid, describes the current phase of the company with two words: "Do better." The company is currently focused on leveraging what is working and making improvements in all areas of its business.

The different teams at VIPKid have different responsibilities and, therefore, share different ideas on how the company can move forward. Though the ideas are diverse, they share a common goal: to make VIPKid as successful as possible for teachers, students, and parents.

Kevyn Klein is always thinking about how to support and grow VIPKid's teaching force. As a next step, she's thinking about how to create more intimate teacher communities where teachers get a chance to make deeper connections with one another. She's thinking about how to encourage teachers to create small groups that meet regularly to discuss VIPKid life. Klein sees the benefit of this as twofold: teachers are supported in their roles, and they are more likely to continue teaching on the platform.

When thinking about these small groups of teachers, Klein said, "If one of them wants to stop teaching on the platform, they all come together and they're like, 'Don't leave. We're a network. We're together. We're friends.'"

Klein is also thinking about how to leverage the communities that VIPKid has created in China and America. On the fun side, she's considering how the teams could come together for a global event. Maybe there's even a way to break a world record for the *Guinness Book of World Records*!

While different teams are thinking about ways to refine and expand the company, there's still a lot of big picture strategizing happening.

"Our mission is to provide the best K-12 education in the world," said Madli Rothla.

Cindy Mi agrees. She believes that online, adaptive classrooms are the future of education and hopes to bring access to these classrooms to all students.

"The world should be more connected," Mi said. "We need to bring teachers, students, and parents working together. This challenge is critical to future innovation."

Traditional education has remained unchanged for thousands of years, but the Internet has ushered in a stage of rapid innovation. A new era of education is upon us, with companies like VIPKid bringing new technologies and approaches to global learning. The benefits of this type of learning are far-reaching, leading to cultural awareness and peace.

VIPKid is in the unique position of having an amazingly successful platform, dedicated teachers, and the ability to scale. It's revolutionizing what online education looks like, building knowledge and cultural connections amongst students and teachers. And it's just getting started.

"They're paving the way for this unique new way of education," said Kim Fortner, "And it's changing the world."

CHAPTER 11
THE LIVING STORY OF TEACHERS

Why I Chose VIPKid: My Journey on the Education Revolution Bandwagon

THI NGHIEM

Thi Nghiem holds a Bachelor of Science degree, completed two years of medical school, and earned a master's degree in public administration in health systems in Massachusetts, USA. He had been a referral programs marketing consultant and private tutor for children of expats in the advance placement and international baccalaureate programs for almost ten years. Thi Nghiem speaks five languages and has traveled to more than twenty countries. Now he is an ESL instructor on the VIPKid platform and is certified through TESOL-VIPKid Advanced and Foundational Programs.

Joining the Online ESL Revolution

I want to start out by saying that VIPKid has provided me with a refreshing perspective on the future of education in relation to the inclusive weaving of cultures and teaching methodologies into its interactive classroom technology. The company's bright sunny-yellow Dino character embodies all the good, and sometimes ideal yet still very attainable, characteristics that we all wish to have for future generations of the world with their continuous spark of curiosity and love of learning in a positive teaching environment.

Working Remotely as a Digital Nomad

VIPKid has provided an additional sustainable source of income for my living while traveling as a digital nomad. I love the freedom to roam the world, the flexibility to open any time slot during available teaching hours in any time zone, and constantly reinvent a better version of myself to stay more focused and organized with my priorities and time management. VIPKid is within itself a creative cloud-based toolbox that I can open anytime and travel to Asia at the touch of my fingertips. I can set up my classroom in any spot that has access to a stable Wi-Fi network and prop up a green screen using ManyCam backgrounds as I move from country to country. Most important, it has opened my mind and eyes to students and their families living in everyday China.

Joining a Camaraderie of Teachers

The teacher community outreach and support are vast and positive throughout multiple social media channels such as YouTube, Twitter, LinkedIn, and Instagram. I learned online TPR, "Total Physical Response," through YouTube videos from VIPKid Brand Ambassadors. It's also provoked my own sparks of creativity to become more animated in and out of the classroom with continuous improv, song/dance, focus on speaking intonation, and simplifying my speaking level to reduce reading speed, resulting in clear phonetic sounds and word pronunciations.

Teaching Methodologies for Different Level Students

Most important, I am always delighted to see my students and their active love for learning English, which sparks rapport building and curious conversations on a one-on-one level. I commit to support higher emotional intelligence when teaching, which revolves around: active listening, self-awareness, and empathy to form healthy relationships while I share their journey to proficient second-language acquisition. Some of my favorite students have been with me since the beginning of my VIPKid journey—since my first contract. Now nearing the end of my second contract, I have earned a loyal following of students who rebook me daily and make me laugh throughout class. Some of my favorite students widely vary in different English-speaking ability, from Pre-VIP to Level 7+.

Engaging Young Students with Fun Digital Props/Rewards

Most of my young Level 2 trial students have both Chinese and English names but are often shy to introduce themselves in the beginning of class, but by the end of our twenty-five-minute one-on-one session, their confidence grows exponentially along with the ability to repeat simple phonics with the aid of my digital props. I love their shocked and surprised facial expressions when they earn digital rewards!

I have one trial student named Xin Er, who also goes by Sophia, whom I converted to one of my most favorite regulars! She needed her mother's assistance in our first few classes together, but after a while, she was excited to sing along independently at the end of every class to the English music videos I would set up as a final reward. She also loves to receive her Princess rewards throughout class, and I often add as bonus digital stickers/animations if she can tell me how many rewards she has received or what color dresses they are wearing.

My regular Level 2s have been very fun to interact with since they often bring their toys to class, develop their own *TPR, and usually need multiple rewards throughout class to stay focused—especially when correcting word sounds and pronunciation multiple times.

I have one very spunky boy, Davy, who loves to jump, play, and draw in our Level 2 classes. He needs a new reward for every slide. Sometimes we shoot bow-and-arrows, make funny sounds, laugh, and sing vocabulary on every slide! He needs ample engagement to stay focused and is more open to repeating mispronounced words several times when he receives a funny digital sticker. One time, he brought a monkey with a banana toy to class, and we ended up eating digital bananas in class as rewards. When he would show me his toy cars, I would find the car rewards, and we'd make zoom and driving race sounds so he never leaves the class screen.

Rapport Building by Sharing Cultural Experiences with Advanced Students

Often during the beginning of Level 4, my students can build a rapport with me on a more personal level, especially during our free talk sessions throughout class. They like to tell me more about their personal lives, hobbies, and favorite foods.

I have a very talented Level 4 student who has told me her mother is an English teacher and she also aspires to be like her mother and myself when she grows up. She always reminds me that English is her favorite subject in school. During every introduction, she excitedly shares her day with me and is the most prepared student I have taught, as she regularly takes notes in class, asks thoughtful questions, and goes beyond the required class homework scope in order to have more Free Talk time. During a particular class, we spoke about going to Disneyland and comparing the one in California to Shanghai. During the math section of the class, she mispronounced "equal," forgetting the "qu-" sound, so I reminded her from my digital prop of a Snow White "queen," which she knows and was able to relate then immediately correct her own mistake. She never forgot this example as we continued into future classes.

What makes me happiest is to see my students grow and bump to the next level if I believe they are bored from the subjects being too easy. In these instances, I reach out to the parents to insist they may reconsider having their son or daughter be more challenged by moving up a level.

Kevin is one of my favorite upper-level students. Most Level 6 classes were often very simple and mundane for him, so he would try to read through each passage so quickly that he sometimes skipped words or did not fully understand the meaning of new vocabulary. So in order to keep him engaged, I often relate other stories, integrate digital props and realia, and make sure to ask thought-provoking questions that allow him to learn actively, not only passively or through mastery of memorization. I believe this teaching style has aided in Kevin's vocabulary expansion for both writing and speaking, thus preparing him for the highest caliber of integration success with English-speaking communities. With my suggestion to his parents, he has been bumped up to Level 7.

Onward—A New Breed of Global Techies

Since teaching on the VIPKid platform, I know there will ultimately be a new generation of jobs not yet created in this new online education industry. I also believe that the young students I'm currently teaching already have increased awareness of future tech growth and will utilize these kinds of Internet platforms profusely to heighten their learning potentials at an undeniably rapid pace. Ideas will be the currency that will fuel their entrepreneurial spirits. My wish in teaching these students is to instill a new seed of understanding of globalization that will result in cultural inclusion, tolerance, peace, and understanding of English language integration worldwide. I hope these connections will allow them to voice their opinions more openly and utilize their knowledge outside the classroom to create a positive social change.

I believe VIPKid will continue to tackle educational hurdles, break tech barriers, and increase its importance in this digital era, ultimately changing the entire landscape for ESL students and teachers, including millennial digital nomads like myself. Again, I'm so humbled and honored to be part of this ground-breaking education revolution where every day I am both a teacher and a student.

Rural Education and My VIPKid Journey
JESSICA

Jessica studied at Nanjing University, in Jiangsu Province, Southeast of China, and now lives in Jianchuan county, Dali, Yunnan province, southwest of China, with her Chinese husband. There, she joined VIPKid's Rural Education Project, and a journey full of love and meaningfulness began.

While sitting in a small hotel room out in Jianchuan county (the southernmost county of the Dali prefecture in Yunnan), I can remember my mom looking at me with a smile and asking, "When you were a kid, did you ever imagine you would end up here?" I answered both truthfully and immediately, "I never imagined once this is where I would be. How lucky am I?" It was my wedding day, and I was the lucky bride of Yancong He. He was on his way to come and pick me up and take me to my new home and my new family.

My new Chinese family was from a very small and very beautiful town in Jianchuan county. Though it is only a few hours away from the tourist hubs of Lijiang and Dali, it feels like a whole different world entirely. I visited Yancong's home for the first time for Chinese New Year in 2015. It was my first time going so far out of a city in China, and I instantly fell in love. I fell in love with the fresh air and warm sunshine that glittered on the rice fields each morning. I fell in love with eating my mother-in-law's home-cooked ham soup while watching the goats jumping up and down as people herded them in and out of the village. Most of all, I fell in love with the brilliant and warm people I met around me who opened their lives, homes, and hearts to help me understand Chinese culture a little bit more.

This was the very same village where my husband had grown up. He told me once that he had run into a foreigner as a child and got very frustrated they couldn't understand him to answer his questions about their lives. His mother told him if he studied English hard, one day he could learn about foreigners and go to their countries, too. As a result of his dedication to English studies, he was able to be the very first student in the area to attend university, to travel abroad for work, and to find himself with a foreign family! Seeing how much English helped him achieve his dreams inside and outside his rural community has always inspired my mission in life to provide easy access to education for rural students.

I had always loved teaching and spent my university years finding as many different excuses to come to China and teach as I could. I had volunteered in elementary schools in Nanjing when I studied at Nanjing University, I found English partners to work with when I was living in Kunming to study at Kunming University, and I even did a summer internship with an international school in

Shanghai to travel to different rural areas to volunteer as a teacher. There was something innately magical about working with passionate kids to help them grow and to feed their joy of learning. It is doubly magical when the students, in fact, believed they didn't like learning and I was able to slyly convince them otherwise and bring their joy alive! I had been bitten by the teacher bug but found myself with families in both the USA and China that limited my ability to take a position in a school.

This struggle to be with both my families while pursuing my love of teaching led me to VIPKid in 2016. I was able to interact with young learners to inspire them on their language journey without limiting my ability to be with either of my loving families. I was enamored with VIPKid's dedication to using a holistic education model that borrowed ideas from both foreign and Chinese teaching practices, using the best of each to provide an engaging curriculum and experience for the students. I felt the joy of teaching come back to me each day as I logged into class to work with my new students.

When I first heard about the Rural Education Project, I knew immediately it was something I needed to be a part of. It was my chance to give my time, passion, and energy back to the same kinds of rural communities that I loved so much. I remember telling my husband about the new program, and his immediate response was "That's amazing! Do you realize if I had had that as a kid how different my life could have been?" I immediately signed up and was overjoyed to join the Rural Education Project team during their second semester.

My first semester with the Rural Education Project, I had a single class. It was such a joy to turn my camera on and see the bright and eager eyes of more than thirty students looking back at me all at once. I worked hard to develop team games to keep them engaged in class, create props and items to help demonstrate the lessons, and practice teaching the lessons over and over throughout the week so my lesson was presented clearly for my students. It was always a lot of work for each class, but it felt so natural and easy, as the classes made me so happy to teach.

When, during my second semester, I was approached by the VIPKid Rural Education Project team to accept three classes in a single semester. I was nervous. Each class would be at a different level and have its own curriculum. In the email, the Rural Education Project team explained that the request came from the local teacher directly, as she had heard through VIPKid that I lived in Dali. Shujie elementary school was near Dali, and they wanted me to work with them. I worked closely with the local teacher and the dedicated Rural Education Project staff to make my schedule work. I worked with grades three, four, and five from the Shujie school. We developed an immediate bond. Classes were full of energy, clear progress goals, and dedication from each of my students.

I would be remiss not to mention that so much of my success that semester came from my partnership with and the dedication of the local teacher, Mia. I often talked with Teacher Mia on WeChat to discuss how I could improve class or what I hoped the students could focus on. I always felt so incredibly lucky to have her working with the students each day, as she was a very dedicated teacher. After each class, she would practice the songs extensively with the students from the curriculum along with all the hand motions. She even helped them to practice the "good-bye" song we sang together at the end of our first class. Imagine my surprise when, before I closed my camera, she asked me to wait so the kids could sing it for me, hand motions and all! I was so moved by the children's studies and Mia's work that I almost cried from happiness. She did a wonderful job working with these students each day to make class fun, engaging, and interactive. She was a wonderful partner.

It was, in fact, Mia who invited me to visit Shujie school toward the end of the semester. We wanted to save it as a surprise for the students. As the semester came to a close, I began to look at my calendar to see when I could drive over for a visit. After sending Mia a message, she told me that the very next day they were having a poetry contest at the school and the students would love for me to visit and watch them. As it happened to be my day off, I decided that, even with short notice, it would be a wonderful time to visit.

That morning, I taught one of my three classes to the students. Unexpectedly, I realized that some of the members of the VIPKid team were sitting in class, as well. The students were their usual, sunny selves, and, as we did for each class, we sang together, danced together, and learned our new lessons. At the end of class, I logged out and jumped in our car. It was time to go! My husband and I drove several hours from our apartment in Dali, and I felt as if I were floating on air. I would finally get to meet each sweet student I had been working with for the whole semester.

When I arrived, the poetry concert was just beginning. I remember seeing a ripple of heads turn toward me and having each new student excitedly whisper, "Teacher Jessica!" and subtly wave at me, trying their best not to distract from the current speaker on stage. I sat down quietly and listened to each of the students in the Shujie school and each of the classes as they presented their material. I have worked with classes from Yunnan to Shanghai, but I had never seen a group of students so committed, diligent, and excited to learn. It truly reaffirmed my mission as an educator to be the best teacher I could be so I could inspire and guide each of my students to fall in love with studying as much as I fell in love with teaching.

When I was introduced, I remember walking excitedly up to the stage, and my heart melted as I heard all of my students scream, "Teacher Jessica!" at the same time.

Some people may think that our relationship was important because of how much my students valued our time together, but, in fact, our relationship as student and teacher was strong because of how much we mutually valued one another. I introduced myself and my husband and talked to the whole school about how very fortunate and humbled I was to work with them. It was my chance out of the classroom to explain to my students how much I hoped that they followed their dreams to study what they love, to explore the world around them, and to never settle for what is expected, but to soar above the mountains around them to find their dreams.

Then, the real fun began. The students were all released to class with a small break, and I was barraged with pictures, hugs, and lots of flowers! By the end, I had so many flowers in my arms and they were so heavy that my husband had to carry some, as well. When the students had found out I was coming, they picked some extra flowers from their own home to give as a gift to me before school started. Just as the people in our Jianchuan hometown had welcomed me and opened their homes to me, so had my students found a way to connect me to their daily lives and homes. I was so moved that I pressed many of the flowers and have kept them with me to this day. As the students would hand me a flower, I would pull them in for a picture, as I wanted to remember each and every one of them and the beautiful gifts they had given me.

I was able to visit each of the three classes I worked with while I visited. It gave me a chance to talk to my students in person using the English we had practiced so hard together. I called on each of the teams by name and practiced asking them questions and demonstrating answers right there in class. We, teacher and student, were finally able to connect without the screen between us. Many of the students came up to me after I talked to them in their class and gave me small notes and pictures they had drawn for me. I was moved to see many of them had practiced using the English we worked on together in class.

Finally, it was time for me to go. We had a long drive back home, and the students had studies to return to. I remember sitting in a nearby town eating dinner with my husband and reading through each of the notes the students had given me. I am sure I looked silly, sitting in the corner of the restaurant leafing through all the multicolored papers, but I couldn't have been any happier than I was at that moment, reading all the thoughts and expressions the students had worked so hard on.

The rest of the semester seemed to fly by faster than I could keep track of. For the final class, the students each prepared one song I had taught them throughout the semester. They sang the songs along with dances to show me all that they had learned. Each team also created a poster to show each of their team members and stood up in front of the classroom to introduce themselves and practice their English. The posters were so incredibly creative! Some drew themselves as princesses or wizards, others working together to build a snowman. Even more heartwarming were the students who included me in their pictures. Yet another reminder that, to them, I was an integral part of their life and team, just as they were to me.

One of my favorite pictures was one that had all boys and one girl named Angela. It was the only team without an even distribution of boys and girls. Angela had worked hard with her team throughout the semester and set herself apart with her bravery in speaking in front of the class. Their poster shows all the boys dressed in their normal clothes in the background and Angela and me standing more prominently at the front, dressed as princesses. It was such a joy to see her express herself in English throughout the semester and, now, to see her express herself through art!

Though the semester ended, the students remained with me in my heart. Working with each of my Rural Education Project classes, including my very special Shujie students, is the exact opportunity I always yearned for to help rural communities grow through helping their children. Children, after all, are the future, and by focusing our efforts on inspiring and guiding this future generation, we build a stronger world for everyone. I had never, ever imagined that through working with the Rural Education Project, I would be able to so strongly influence so many lives, nor had I imagined those same students would inspire *me* as a teacher to continue to grow and push myself to new limits. So, as I sat on my wedding day preparing to marry my now-husband, looking at my mother and listening as she asked me, "Did you ever imagine you would end up here?" I smiled at her, thinking of my Shujie students and all the rural students I had met. I thought of all the amazing places these students would take themselves through their own personal perseverance and language studies, places they perhaps could not even imagine yet themselves. I smiled and said to my mother, "I never imagined once this is where I would be. How lucky am I?"

Believing Without Seeing (the Definition of Faith)

ROWENA GESICK (TEACHER ROWENA)

I am a proud mom, wife, and teacher. I've been fortunate enough to have lived in some of the most beautiful places in the world. Some of these places include the Philippines, Hawaii, Germany, Las Vegas, Georgia, and Virginia. Growing up and living in these different parts of the world has given me a rich experience of other cultures and the

*opportunity to meet some amazing people. I think
this is such a benefit for an ESL teacher. I've lived in
coastal Virginia for the past two decades and enjoy
my life as a sports mom, wife, and teacher. I love to
write and to travel, I enjoy the outdoors, cooking,
and most especially watching my kids play sports
and pursue their passions. I'm also a mom to an epilepsy warrior and a big advocate on epilepsy awareness and education. My husband, two kids, and two
German shepherds make up our family. They are my
inspiration and make me proud to be a mom.*

At a young age, I knew I wanted to pursue a career in
education. This was because of an event in history that
changed the world. I was in elementary school when the
lives of seven brave souls were lost after the space shuttle
Challenger took off into the Florida sky. One of those
souls would have been the first teacher in space. That
remarkable teacher inspired the world with her ambitious soul and brave heart. It was then I knew I wanted to
be a teacher. I shared this goal with my big sister, who
was always my biggest supporter. She passed away a few
years ago of breast cancer, but I know she continues to
cheer me on from heaven along with my dad, who was an
educator for over thirty-five years. I know they both
would have been so proud that I've been fortunate
enough to have been given that opportunity.

In 2015, my youngest daughter was diagnosed with

epilepsy. She was seven years old when she suffered her first seizure. That moment changed our lives forever. We knew that the road ahead would not be easy. My husband and I never showed our children the fears, uncertainty, or worries we felt. I thought about my sister and her battle with breast cancer. And every once in a while, I would catch a glimpse of her in my daughter's eyes. The eyes of a young child with so much will and determination.

At the time of my daughter's diagnosis, I was working for the school system teaching at a local high school, as well as in school full-time, finally getting the opportunity to complete my bachelor's degree. The plan was to start immediately working on my master's degree. However, I put those aspirations on hold and made the decision to go on family medical leave. I knew the road ahead for my daughter and our family would require more time off from work for doctor's appointments, clinic visits, lab testing, and many meetings with her medical and educational team.

That first year included some of the most difficult times we've ever had to go through as a family. My youngest child was sick and did not know how to handle the unpredictable storm known as epilepsy. She also struggled at school, and it was heartbreaking to see that a strong-willed child with so much intelligence and will was now being affected by a condition that observes no boundaries when it hits you. As a teacher and most especially a mother, this weighed heavily on my heart. But when you have a child who refuses to be defined by an illness or medical condition, when you have a child with such strength and will and determination, it's hard to give up. Maya Angelou once wrote, "You may encounter many defeats, but you must not be defeated." I often remind both of my children this important take on life.

Those next two years were spent at many doctor's appointments, many unpredictable doctor's visits, a good number of medical tests, many phone calls from the school nurses, and many days left wondering why life takes such unexpected turns. Those next two years, my faith was definitely tested. My daughter continued to struggle with school and sports because of the side effects of epilepsy and medications. But she continued to fight bravely and then openly to reach her goals. She continued to play soccer and basketball no matter how much she was physically in pain; no matter how much her body was fatigued, she kept on moving. And I could not be prouder of her strength and determination. I could not have been prouder of the faith that she has in herself and in me. In the midst of this, my oldest daughter, who had done so well with middle school athletics, developed a painful foot condition in the middle of her track season. This foot condition sidelined her, and we were told she might not be able to run track. After months working with two doctors and six months of physical therapy later, she regained her strength and was able to compete in both soccer and track again. I remember a moment in sixth grade when I asked her, "What would you do if you fell during a race?" She told me then, "Mommy, I'd get up and keep running." And from that moment on, that was our family motto. We live by it and remind one another of this statement during tough times.

I eventually resigned from my teaching position with the school system, and I put my master's degree on hold because my main focus was and is my family. Not having that second income I brought in from my teaching job hurt us financially. However, we knew that the sacrifice we made was bigger than finances. And we kept our faith that things would get better. Then the summer of 2017, I found VIPKid, or VIPKid found me. And by December of that year, I signed my first contract and taught my first classes.

Like my family's challenges, the start of my VIPKid journey was also a challenge. But I had faith that this was going to be something special. I knew that I had to work harder and smarter to build a solid foundation. As I took more workshops and discovered more videos online posted by teachers and also joined the online community, my bookings began to increase. One particular teacher was so encouraging and showed so much faith in me when I attempted my certification for TOEFL Primary. I knew that having that certification would be significant. The exam was tough, and it did take a couple of attempts to finally pass it. It was discouraging, but that particular teacher would not allow me to give up. We were strangers and only knew each other through VIPKid and our online community, but it was really amazing to have that kind of online support. I am forever grateful to her for it, and she continues to believe in me, uplifting me and others.

In June of 2018, I started my second contract with VIPKid. That summer really made a huge difference in my progress. I was certified in more levels, and I earned certifications in supplementary classes. It was all due to our online VIPKid community. In particular, the TOEFL Primary certification was of significant value. I met a very special student, Bobby from Nanjing, China. He is an extraordinary boy who genuinely loves to learn. He made such an impression on me during that first TOEFL Primary class. It just felt natural to speak to him about his day, his interests, and his family. I can remember one of his parents' first feedbacks to me stating how they appreciated that I had a genuine interest in the life of their son in China and how it was important to them for him to be able to practice conversational English. We started with TOEFL Primary classes once a week, then eventually major course classes, and then he was on my regular schedule twice a week.

TOEFL Primary also led me to another wonderful student, Julie. In one of our first classes, she shared with me that she played a stringed instrument but did not know the name of it in English. She showed me what it was, and I did my research on it. I learned it is called a "Chinese Zither." The next class we had together as we were saying our good-byes, she said, "Teacher, wait one moment." She turned the camera to show me her Chinese Zither. I could tell she wanted to play it for me, and since we had a little time, I asked her to. The joy in her face and in how she played was so beautiful. She took my breath away with the music. I will never forget that moment. Another unforgettable moment is finding out that Bobby and Julie actually know each other! They both told me they live in Nanjing and attend school together. It was such a surprise and so special to know that all this time I had been their teacher, they knew each other.

Bobby continues to be a model student and continues to connect me with other students. In particular, Tony, a newer student whom I have had the pleasure of teaching for a couple of months, told me he lived in Nanjing, and I mentioned that I also had students there. To my surprise, he asked, "Is it Bobby?" His mom then proceeded to tell me that Bobby's mom had recommended me to Tony's mom. I was overjoyed! And this special online connection continues. A student named Shirley who is also a newer student mentioned that she knew Bobby and Tony. Her mom then showed me a picture of the three of them together. I am just completely amazed and grateful that these connections were made simply from me working to get my TOEFL Primary certificate last summer and having a fellow teacher genuinely believe in me. I am so humbled and inspired by these connections.

I wish I could mention every student that I have taught who continues to inspire me. There's the most amazing four-year-old boy, Andrew, with a million-dollar smile, whose abilities and focus are beyond his years. And whose mom has offered to be my tour guide in Shanghai. She also recommended me to her longtime friend who has a four-year-old daughter named Xinxin, with a shy smile and the sweetest laugh. And Vito, a talented student who took some time to connect with me, but whom I now look forward to teaching twice a week. Coco, with her carefree spirit and loving personality, who greets me with a favorite toy at the start of each class. Peter, with his bright eyes and supportive mom who offered me an open invitation to their home and said she'll make me homemade dumplings. And Gina, the first student to draw a picture of me for her homework project. These are just some of the amazing students whom I have the opportunity to teach and who continue to amaze and inspire me.

In December of 2018, I was fortunate enough to sign my third contract with VIPKid. I was overwhelmed with emotion. It was everything that I had hoped and worked for. I was so overwhelmed with gratitude and relief. I always knew I had a passion for teaching, and I knew that I needed to show that passion to my students and through my teaching. And here I am, in my third contract with a business teaching the best, most ambitious, beautiful bright-eyed children across the world.

VIPKid is my family's saving grace. We are so much better not just financially, but our overall spirit and faith are still intact. My children are growing and doing well. My oldest daughter overcame her foot condition and continues to excel in soccer and track. She ran at the New Balance Indoor Nationals this past March in New York City. And most recently, she ran at the prestigious Penn Relays in Philadelphia, and colleges are taking notice. I am a proud sports mom. My youngest continues to fight epilepsy each day. She has learned that she is a warrior. Both of my children know that VIPKid has given me the advantages and benefits of working from home and the ability to take care of my family in all aspects. I am there for major events and milestones because I have the freedom to make my own schedule. I am able to help my husband financially and to support our children's goals and aspirations. VIPKid has given me the opportunity to pursue and live my dream and passion of being a teacher.

I am fortunate that I have so many people who count on me and believe in me from China to the United States. I am so inspired by the children I teach; I admire their will and courage to learn a new language. I am inspired by my fellow teachers, who spread so much joy and who genuinely care about making this world a better place by the faith and sacrifices they make. I'm proud of my own children, who have taught me that a little faith goes a long way and when we fall, we get back up and keep running.

The story of Rowena's daughter is accessible on YouTube in My Epilepsy Hero campaign https://youtu.be/QEqlk7kd4PU.

Travel, Love, and VIPKid

KATHLEEN SCHMIDT

Kathleen Schmidt, Kathy for short, was born in South Africa and lived there during her formative years before immigrating to the United States with her family. During the last two decades, this forty-one-year-old teacher, who has a bachelor's degree in environmental education and over six years teaching in that field, as well as spending three years teaching ESL to K-12 students in Seoul, South Korea, has traveled to over seventy countries and had a wide range of experiences. These vary from teaching English abroad, to working as a scuba diving instructor in Thailand, to volunteering in the Amazon, and even working on cruise ships. Kathy met her husband, who is from North Macedonia, in May 2016. They moved to North Macedonia, and she started her story with VIPKid.

This is a story of travel and discovery, but, most important, of a love that led to a new country and how that love led me to finding VIPKid. It is about the journey I have made through more than seventy countries to end full circle doing something I love, with a man I adore and helping the parents who have encouraged me.

I found VIPKid by pure luck! Sometimes I wonder if all my past experiences were leading me toward a lifestyle where I would need to find work that allowed schedule flexibility and offered great pay. This is my VIPKid story.

Throughout all my many, varied life experiences, my time teaching English in Seoul, South Korea, was one of my favorites and something I always hoped to go back to in some form. However, opportunities continued to knock, and I spent many years traveling the world, including working as a scuba diving instructor in Thailand and as a cruise ship worker sailing everywhere from the Caribbean to Ukraine. I even spent two months volunteering in the Ecuadorian Amazon at a wildlife refuge and a month volunteering on the island of Tonga with an animal welfare group. However, I knew that eventually I would need to return to the US long term, as my mother suffers from multiple sclerosis and would need me to be there to help care for her at some point. It looked as if returning to teaching ESL abroad was not to be. Or so I thought.

In 2016, while working on a ship in Norway, I walked into one of the ship bars and met a waiter named Dane from the Republic of North Macedonia. According to him, it was love at first sight, but it took me a few weeks to get the memo. A few months later, we disembarked and moved to his home country of North Macedonia to get married. Here I was, living in a new country, faced with a language that was completely different from any language I had been exposed to, and at a complete loss as to what to do for work.

Dane was working as a chef at a local restaurant, but salaries in Macedonia are very low, and he was bringing home barely $250 a month, just enough for basic expenses. This is when I started to search for something I could do while struggling with not speaking the Macedonian language and without needing to have a work visa. I signed up to teach on the VIPKid platform. I opened my first class in December of 2016 and met Julien, my very first student. He was awesome and gave me the confidence I needed. He also continued to book me for the next two years. Back then, VIPKid was still new, and teaching English online was something no one had really heard of, so there were many who doubted the validity of "teaching English online." Of course, this mindset has changed quite a bit since then.

December in Macedonia can be rather chilly, and of course Murphy's law had something to say about me needing stable Internet connection and soon blasted us with a winter like no other. We had three months of temperatures below -20F, one of the coldest Macedonia had experienced in ten years. The cold resulted in no Internet at our house and no technician willing to brave the frigid temperatures to fix it. So Dane and I made a plan, and I spent most of my first few weeks teaching from the store room of the local coffee shop, sitting on drink crates and using the freezer as my table. While the store room was warmer than outside, I still needed gloves, a hat, and a scarf. On top of that, I was still getting my footing to figure out my teaching style. My winter wardrobe made such an impression on my students that the first time I was back home sitting next to our nice warm wood stove, they all asked why I looked so warm that day compared to the classes before.

Living in Europe meant I could teach during the day, 9 a.m. to 2 p.m., but unfortunately, Dane's chef job had him out of the house from 3 p.m. to 2 a.m., so we had little time together for two months. Eventually, we decided that I would add a few more classes to my schedule and he would become my own personal chef at home, responsible for lunch and coffee. Pretty sweet setup if you ask me. All I had to do was open the door, nod my head, and coffee would arrive on the desk while I was teaching. Lunch was always ready and always delicious! It also meant that we were earning way more than before, as I was able to get booked enough to bring in almost $2,000 a month. In Macedonia, this is considered a fortune, and so with this money we started some much needed renovations to his mother's house where we lived, redoing everything from the bathroom to the front door, and still managed to save some on the side.

One of the greatest benefits to teaching with VIPKid is the ability to create my own schedule. This meant we could go on trips and explore many parts of Europe. We even did a monthlong drive through the Balkans and visited six different countries as a belated honeymoon. These days, there is a whole community dedicated to travel and teaching, which also means it brings the world into the classrooms when teachers tell their students where they are and what they see.

After two years of teaching 9 a.m. to 3 p.m. in Europe, we moved back to the United States. My parents live in Eugene, Oregon, and my mom's multiple sclerosis had progressed to the point where help was needed. Moving to the West Coast meant a completely different, completely opposite schedule; all available teaching times were between 6 p.m. and 6 a.m., depending on daylight savings time. If that weren't enough, my classroom is in the bedroom next door to my parents' room, so this all meant a total upheaval of my schedule. I went from teaching around fifty classes a week to barely twelve. Luckily, I have managed to teach some amazing kids who are super supportive and always try to book me. These days, I teach 6 p.m. to 9 p.m. on Fridays and Saturdays and plan to open more classes during the summer.

Something I love more than anything is sharing my travel adventures with my students. There is Mango, an eight-year-old girl, whom I have taught for two years now. She told me, "Teacher, when I turn twenty-one, I will go to Australia!" There was no hesitation, she has decided she is going. So I shared pictures and stories with her about my visit to Australia; I also sent a read-along book to her, which she practices reading before class and reads me a new page each lesson. We are now able to have more and more discussions about different things in Australia; she is learning something new every lesson and preparing herself for her trip there.

Then there is a little girl named Coco. She and her mom loved seeing my pictures. At the end of every lesson, I would give her a choice of what country she wanted to "visit" with me in the next class, and I would use the pictures from travels in that country as a reward. They were always so amazed and excited to see them.

In all, VIPKid has allowed me to continue doing something I love: teaching. It has provided the finances for a monthlong honeymoon trip through the Balkans; allowed us to make much-needed, major renovations to my mother-in-law's house; financed our move back to the US; and now covers our main expenses. The VIPKid team in China and in San Francisco are supportive and always willing to lend a hand or give a suggestion.

Being back in the US has also meant meeting more teachers (there were no other teachers in Macedonia, although I did manage to recruit one). Here in Eugene,

Oregon, I am the teacher who has been teaching on the VIPKid platform the longest and am always willing to share my experiences and give support and encouragement. I have even started hosting events: the first was a fundraiser for the Multiple Sclerosis (MS) Society. In just a couple of weeks, the VIPKid community in Eugene raised $350. Unfortunately, the main event was rained out, but that didn't get us down. We still met for coffee and pictures and are planning to meet up at a rescheduled time to show our support for my mom and others while raising awareness for a cure for MS.

When I started over two-and-a-half years ago, VIPKid was like a bright, shiny new penny, with only a couple of thousand teachers and students. Now, it is leading the way on a whole new form of education, while bringing two cultures closer and instilling understanding into worlds neither students nor teachers could ever before imagine. VIPKid has become a world leader in online education and is expanding and growing above and beyond any and all of my expectations. VIPKid is a true pioneer in the field of online education, and I can't imagine my life without it.

Discovering Myself from across the World

DONNA N. DUNBAR

This isn't a story about tragedy. This is a story about change and hope.

In 2013, I died.

I had my first heart attack; I died on the table.

Then I had my knees replaced and learned to walk all over again.

It was not a good year for me at all. I had very little hope for a future and had no idea what my life had in store. I was too unhealthy to work because I could barely walk and was extremely overweight.

In 2014, we started a new life, moving to a new state. I had two high-school-aged boys still living at home. Like many parents, I had devoted my life to my children and their education. I stayed at home with them to teach because I felt that not only did it give me more time with them, but it also allowed me to give them the best possible education.

Every year, I asked my boys if they wanted to continue learning at home or go to public school. In 2015, for the first time, they said they wanted to attend public school.

I considered going back into the classroom, but I did not have a teaching contract for a new state, and I really wasn't in the best of health to take on full days of teaching. What could I do to fill in my days and time? My life had been my children, but the children were growing up.

I longed to get back to teaching. My life felt empty! I began to change my life, because I was tired of being extremely unhealthy and of doing nothing. I started looking at new things to do. I changed my living habits to become a better me.

At the same time, I scanned the Internet looking for jobs that would fit my lifestyle. I love children and working with children, and I missed it. So I thought about a few writing jobs that were available; however, they just didn't resonate well with me. I looked on this website that sells used things that also had a small job section. Occasionally, there was an ad looking for teachers from a Chinese company by the name of VIPKid.

I thought it had to be a scam. How much money were they going to charge me to say I could teach? What if I worked and they never paid me? I had a lot of questions. It just didn't seem real.

So, with these questions, I started researching the company. At that time, there wasn't a lot of information about VIPKid. It was a relatively new company. Despite this, I decided to apply. What was the worst that could happen?

In the fall of 2015, I had my demo lesson. I had several stitches in my stomach and couldn't stand up straight when I walked. It was a good thing it was online and I could sit through the demo lesson.

I signed my first contract with VIPKid moving into a more positive position in life. I was starting to go in a new direction that led away from the negativity of my past toward discovering who I was presently. I was feeling as if my life had value again. Through VIPKid, I renewed my love for teaching, and my love for living.

VIPKid soon turned out to be much more than a job. It became my gateway to self-discovery. I quickly made new friends from all over the world. I met students that have a special place in my heart.

For the first time in my life, I was getting up early and loving every minute of it. My family said I sounded happy when teaching. And I was! I learned so much about my students, and I shared some things about myself, as well. The children from across the globe were interested in who I was, and I think it helped them progress in English, too.

When they asked me how old I was, my standard answer was "I'll tell you, but it's a secret. Shhh . . ." Then I would take my whiteboard and write "988." Many eyes would get big, but one student told me, "If you are 988, then I am 1,000." I was making connections. My heart grew stronger.

I started to look forward to waking up in the morning. My health was improving, and I was losing weight and living a healthier life. In truth, I was becoming a new me. I was able to travel and meet people inside the company and grow professionally in ways I never dreamed would be possible. With VIPKid, I don't have just students, but I have gained an extended family.

In 2017, a student asked me whether I had a plan to go to China. I laughed. I told him that I didn't have plans to do so, but that it would be wonderful if I could. But the surprise was that a few weeks later, I received an email saying I had won a trip to China. I was stunned.

While on the plane, I still couldn't believe that I was headed to Beijing. I had the honor of meeting a wonderful group of teachers and some of the people I had spoken with at the Beijing offices in person and not just via chat or online. It was such an honor, and I found out why VIPKid was such a spectacular place. They have some of the most talented, intelligent, and kind people I had ever met in my life.

My impression of China completely changed. What you read in books and see on the news is nothing like the experience of being there. The Chinese people were so kind, polite, and helpful. The country was beautiful. I walked the Great Wall and even challenged a fellow teacher to a plank challenge. We planked on the steps of the Great Wall for three minutes, declaring a tie. The Great Wall was massive and went on as far as the eyes could see. It was also full of mazes, and I was the only teacher in the group who got lost.

The mountains were majestic. I truly felt I was on top of the world. At the entrance, I saw two children trying to see a dog under a small building. She had puppies. I stopped to buy the mother dog some food and water. I wanted to give back to China because it had already given so much to me.

Thanks to this work, I had the opportunity and pleasure to meet and spend time with some of my wonderful students. I learned that even a grandmother could still beat a nine-year-old boy at a wall squat, much to his surprise. I also learned that a five-year-old Chinese boy can still love you as much in person as he does online. And that the girl that you held dear in class was just as dear in person, and I learned she was a dancer and an actress.

My students had so much talent and so many dimensions that you cannot possibly learn in just one twenty-five-minute online class. They are truly remarkable young people.

The parents took me to dinner and even sat with me for three hours at the airport when my flight was delayed, refusing to leave me because they wanted to ensure I was safe and made the plane on time. They weren't strangers: they were friends, and I felt loved. I left a piece of my heart in China.

Some may have come to VIPKid to earn extra money or to be able to work in an arena where they could travel across the globe, but for me, it has been so much more. VIPKid has given me a new lease on life, and a new joy of discovering what the rest of my life has to offer. In other words, life with VIPKid is an adventure for the mind and the heart. If anyone were to ask me what I would want to do with my life, my answer is this: "I'm doing it!"

From Exhausted to Exhilarated: My VIPKid Journey

DEANNA CLARK

Teacher Deanna lives in Holcomb, Kansas, in the United States, with her husband and two children. She has been a teacher in a brick-and-mortar classroom since 2004 and has been teaching on the VIPKid platform for a short time. Deanna grew up in a family of educators, and she knew that she would follow in her parents' footsteps. She received her bachelor's degree in English from Fort Hays State University in the spring of 2004 and began teaching shortly after.

I have taught in five different schools over the span of my career, from small, rural schools with fewer than fifty kids in four grades, to larger schools, with more than eight hundred students in two grades. I have taught everything from middle-school English to college composition classes, but I currently teach seventh-grade students in Garden City, Kansas. I've enjoyed English language and literature since I was a child, and my love for literature has continued to blossom into adulthood. I read continuously and hope to pass that love for literature on to my students in the classroom.

Teaching has always been my passion, and seeing the light in my students' eyes when they grasp a concept for the first time has always been my inspiration to continue teaching. Until very recently, that light has been enough to keep me going in the classroom, working to help each of my students reach new heights in learning. But just recently, things have begun to change for me in the classroom—both in my attitude and my passion for educating. There are several reasons why.

Over the past fifteen years, education has not been a high priority in the state of Kansas, where I live and teach. More and more funding has been cut from the classroom each year. Teachers have been expected to do more, but we have been given less. I see kids with all kinds of issues, not only educational, but also emotional, social, and physical problems. Kids are coming to my classroom who do not have parents or adults in their lives to help them succeed. Some of my twelve-year-old students are even expected to cook, clean, and care for their younger brothers and sisters. Homework isn't a high priority when you have to put dinner on the table. In the past few years, I've also seen a rise in emotional turmoil among my students. Today, it's a constant battle, but the kids have kept me going. I began teaching because of my love of literature. I've continued to teach because of my love for my students. What other business opportunity lets you mold young minds—minds that will eventually lead the world? What other job lets you touch the lives of thousands of people, not just a handful? What other job gives you the gratification of having someone you've helped through both educational and emotional struggles come back to tell you you've made a difference? I can't think of any besides teaching.

Unfortunately, things changed for me in the past couple of years. I became tired. Tired of state-mandated testing. Tired of discipline issues. Tired of grading hundreds of essays and assignments. Tired of spending hours of my time lesson planning. Tired of jumping through hoops created by legislators with no idea what it's like to be in the classroom. Tired of depleting my budget at home so I could give my students what their parents could not afford to give them.

I love my students, and I love what I get to do every single day. I love watching my students grow and learn right before my eyes. However, I'm tired of not having the resources to help my students. I have spent thousands of dollars out of my own pocket to buy basic supplies for my students to use in the classroom—pencils, paper, books, etc. I've also used my own money to purchase food for my students—being hungry doesn't help students trying to learn.

This year I began a snack program at my school called the "Hawk Pantry," named after our school mascot. I created several Donors Choose projects and started three Facebook fundraisers to raise money to buy healthy snacks for our students to have, free of charge. It's been a resounding success. We have fed more than eight hundred students for nearly a full school year, all on donations and fundraising efforts. The difference in student performance has been astounding. But even that hasn't been enough to keep my passion in full force. I've been searching for a way to light the fire I once had for teaching. While the Hawk Pantry gave me an exciting opportunity to help my students in new ways, it also nearly derailed my teaching career altogether. I began researching nonprofits, thinking maybe a complete career change was what I needed to get excited again. However, I was unable to find anything that really sparked my interest, so I decided to stay in the classroom.

Along with teaching full-time, I've worked many side jobs during my career. Teacher salaries in the United States, and Kansas in particular, are low. At the end of a day in the classroom, the last thing I want to do is go put on my Walmart smock, or go weigh trucks at the local grain elevator, but for many years, that's exactly what I've done. I've always been on the lookout for a way to use my teaching skills outside of my classroom, but those opportunities have been few and far between, and hard to come by.

On a whim, I decided to sign up for VIPKid. I had seen the ads on Facebook and had even gone so far as to fill out my preliminary information before getting sidetracked several months earlier. Having the opportunity to teach from home seemed out of reach. How was it even possible? What if my style of teaching didn't translate to the online world? What if no one booked me? Wouldn't it cost a lot of money to get started? Was I as animated as all of the teachers in the example videos I'd seen? Even with these questions weighing on my mind, I decided to give it a chance. I really didn't think I would pass the first steps because I'd never taught online before. But when I passed the initial phase with flying colors, I decided to try the mock classes. I didn't expect to pass them on the first try either, but I passed the three I took immediately. I had seen all those advertisements for VIPKid on Facebook, but I'd never met anyone who had actually gone through the sign-up process or worked on the platform. Once I began passing through the sign-up process, I began to do more research. It sounded like a dream come true!

Signing up for VIPKid has been life-changing for me. I have found a new energy and passion for teaching for the first time in five years. That fire I was struggling to find in the classroom is now burning with vigor. I get to spend my mornings and weekends with children in China, and their energy rejuvenates me to work with my students here in America. My American students love to hear about the students I'm working with overseas! They are constantly asking if I've taught any kids like them, and what kinds of things my Chinese students are learning.

What I've found is that kids are the same worldwide. While the attitude toward education may be different in America and China, kids are kids, no matter where they are. Seeing my students on both sides of the world learn and blossom has been a truly gratifying experience, and one I wouldn't trade for anything. I truly believe that my viewpoint as an American teacher will help open new worlds for my students from around the globe. I have had experiences that they haven't had, and vice versa. Not only are they learning from me, but I am also learning from them. By working together, my students and I can foster a spirit of understanding and acceptance for those who are different from us. I believe that a well-rounded education must include not only rigor and relevance, but respect and relationships. If I can teach my students to understand and accept those who are different than they, I am doing my job. Yes, grammar, spelling, and math are important, but so are kindness, caring, and empathy.

Teaching on the VIPKid platform has not only changed my world financially, it has also changed how I teach in my brick-and-mortar classroom. I catch myself using more animation and TPR. Where before I might have had moments where my voice became monotonous, now I can hear more animation and excitement as we tackle new topics. My students have noticed a difference too, and they have been so excited to talk to me about my students from across the world. VIPKid isn't just helping my students online; it is helping my students in the United States, as well.

The lessons are straightforward, the kids are amazing, and the support is outstanding. I have created new friendships with other teachers, and I've even convinced several of my American teacher friends to sign up to teach on the VIPKid platform. Two are almost ready to start teaching! They've seen the enthusiasm and excitement I've found, and they too want the opportunity to rekindle their love of teaching.

Thank you, VIPKid, for giving me back my passion for teaching! Thank you for reminding me that what I do every single day makes a difference in the lives of many. Thank you for reminding me that I can be the teacher I was when I graduated from college with hopes and dreams and excitement. And thank you for giving me the opportunity to touch the lives of students across the globe. I can't wait to see what my VIPKid future holds.

Embracing Change

JEN MOSKALL

Jen Moskall lives with her husband in Florida. She received her bachelor's degree in social work from the University of Michigan, and a master's degree in early childhood special education from Florida State University. After she received her elementary education certification, she became a teacher. However, in 2017, she quit that tenured, retirement-guaranteed state job, and began teaching on the VIPKid platform. This story will explain why.

My story begins like that of most teachers. I have a passion for working with children and helping others, which led me to a career as a teacher. Unfortunately, like too many teachers, I was burnt out after only thirteen years. My overall health was declining too rapidly for an early forty-year-old. Since joining VIPKid, my mental, physical, and emotional health are all thriving!

My name is Jen Moskall. My husband and I live in sunny southwest Florida with our two dogs. I love everything about being outdoors in the warm sunshine. I love to swim, practice yoga, ride my bike, and fish. Before I became a teacher, I worked as a case manager for developmentally disabled adults and as a school social worker for Head Start. I then transitioned to teaching after I earned my elementary education certification degree. As a teacher, I worked at a brick-and-mortar school for nine years. During those nine years, I worked as a kindergarten teacher, reading coach, and a special education resource teacher. I loved working with small groups of students to help them become the best students they could be, but the stress of a large caseload led me to find other opportunities. I then became a hospital-home-bound special education teacher for students who were too ill to physically attend school. It was during this time that I began teaching with VIPKid. For eight months, I worked my full-time job and taught on the VIPKid platform. It was not easy, but I saw an opportunity with VIPKid that could change not only my career trajectory, but also improve my mental, emotional, and physical health.

I have always loved spending time with young children, and teaching was a great way to access my passion for children and helping others. Regrettably, there were no teaching positions in Michigan, where my husband and I were born and raised. My husband and I decided to take a leap of faith, along with over-the-phone job offers, and move to Florida. I loved teaching! I loved being a teacher! Each day was a challenge and exhausting, but the students' kindness and curious minds always put a smile on my face and in my heart.

Unfortunately, after more than a decade of teaching, I was overly stressed, unhealthy, and overall burnt out. In August of 2017, my health became a serious issue. My doctor informed me that I had a mass in my pelvis that needed to be removed as soon as possible. I had been having multiple health issues over the past year, and I was finally receiving a diagnosis. Along with this diagnosis came many tests and eventually surgery. The doctors and surgeons were not sure what the mass was. Was it a tumor, a cyst, or, worse yet, cancer? The only way to find out was to have surgery. To say I was worried and scared is an understatement. For ten days, I lived with the thought and fear that I might have cancer. These are not healthy or productive thoughts, and they overwhelmed most of my days. Once surgery was complete, it was determined that I did *not* have cancer, but that I did have a tumor and two cysts, all of which were removed. This was a day of pure happiness and relief. I did *not* have cancer!

It was during my nine-week recovery that I began reflecting on my career, my health, and my overall happiness. Being told that I might have cancer changed my perspective on how I wanted to continue my career and live my life. It did not take me long to determine that I needed a change, specifically a career change. I needed to reevaluate my career choices and get back to finding joy in teaching children because the thought of going back to my teaching position filled me with anxiety and sadness. I love teaching children, but I had no idea how to renew my passion without the stress of a state teaching job. I needed to find a job that would allow me to teach children in an inspiring and engaging format. I made a promise to myself: I would not continue to live the unhealthy, unhappy life I had been living. I would spend my medical leave reflecting on, searching for, and planning my future happiness.

After the first week on medical leave, I was going stir-crazy! I was healthy enough to take care of myself, but not healthy enough to drive or go anywhere. I typically taught special education primary students, which means there is no downtime and it is go, go, go all day long. I needed something that I could do from home and keep my brain active and engaged without putting stress on my recovering body. I also was hoping to find something that would extend past my medical leave, in hopes of finding a change I desperately needed. While reflecting on and searching for options online, I remembered my best friend talking about being an online teacher.

My best friend had started teaching with VIPKid the previous year. She had been nothing but positive when talking about this company and how much she enjoyed teaching online. Now that I had nine weeks of nothing to do, I was going to look at VIPKid for myself. As I began asking my friend questions, reading online reviews, and watching countless videos, I realized I may have found the change I needed. Just before Hurricane Irma hit, I submitted my basic information and spoke to the company. But then Irma struck Southwest Florida. My family and I were safe, and our home suffered very minor damage, but we were without power for twelve days. This put my sign-up process on hold. I was anxious and nervous. I felt that if I didn't pass my mock class soon, my dream of quitting my state teaching position would be crushed. My anxiousness and nerves were for nothing, because thanks to some wonderful coaches and my referring teacher, I signed a contract on September 10, 2017.

Now that power was restored, I began preparing my classroom and taking as many workshops as I could to prepare myself for teaching on the platform.

Before my nine-week medical leave was complete, I was teaching four to six classes, five or six mornings a week. I fell in love! My passion for teaching had been restored. These sweet, curious, yet sometimes serious faces of children in China were quickly capturing my heart. I had been struggling with teaching for the public-school system for various reasons for quite some time.

As a public-school teacher, I often felt that my hands were tied, that I could not offer my time and resources to students in a way that they needed. I believe that teachers and students both perform at their best when there is not only a love for teaching, but also a love for learning. Even when I was doing my best to maintain my love for teaching, my students were not finding a love for learning. They were stressed and discouraged by the academic demands placed on them at such a young age. Requiring primary-aged students to read and complete multiple-choice and short-answer math and/or reading assignments and assessments is challenging, but asking a child with a learning disability to complete these tasks is disheartening to not only the teacher, but the student, as well. The time spent on these difficult tasks could be spent nurturing the students' learning abilities; instead, it focuses on the students' deficits and enhances their frustration.

When I teach my VIPKid students, I feel not only

my passion for teaching, but also the students' passion for learning English. For the past several months, I have been Emily's teacher. When I first met Emily, she was a quiet and serious Level 2 student who was often frustrated. She was having difficulty with grammar rules, pronunciation, and fluency. Forty-four classes and 110 minutes later, Emily is successfully working her way through Unit 3. Yes, Emily still struggles with pronunciation and even grammar rules at times, but she no longer is the same quiet and serious student I met forty-four classes ago. Now she comes to class, smiling and usually bouncing in her chair, excited to learn with me. Emily tells me about her brother, her favorite toys, and her classmates. The best part of every class, though, is that our classes always end with our hands in a heart shape saying I love you. Even after my camera is off, I can still see and hear Emily telling me how much she loves me and that she will see me again soon.

VIPKid is a much-needed breath of fresh air. After thirteen years of working for the public-school system, I made one of the most difficult decisions I have ever had to make: I quit my tenured, retirement-guaranteed state job. It felt like a weight had been lifted from my shoulders. I could now teach in an inspiring and engaging format that I had dreamed of.

I now teach an average of sixty classes per week. I also teach with the one-to-many programs and find new teachers whenever I can. I have regular students whom I look forward to seeing each week. I have built relationships with not only my students, but their families, as well. My VIPKid students have become a part of my family. I have welcomed them into my home and my heart. I wake up most mornings thinking this is too good to be true. I get to teach children on the other side of the world how to speak my own language. These students have brought so much joy and happiness to my life and have helped me find my passion for teaching again. Since leaving my state teaching position, my health has greatly improved. I now have time to enjoy swimming and yoga daily. I no longer have stress or stress-induced headaches. I am a genuinely happy person. Thank you to VIPKid and to all my amazing students and their families! I am thankful for the VIPKid family each and every day!

I'm More a Mother than a Teacher

ZHAO DONGMEI

Zhao Dongmei is a book editor and a freelance writer. Her daughter is learning English through VIPKid and enjoys learning on the platform. After seeing her daughter's experience, Zhao Dongmei herself became a Lingo Bus teacher and has learned to play the role of teacher in addition to mother. It is this transition that helps her to give all her love and obligation to her students.

It was lucky for me to find Lingo Bus, which has helped me reexamine myself as a mother and teacher, and how I could both love and teach all of my kids—both at home and in the classroom.

My daughter Cynthia is just five years old and has been learning English online with VIPKid for one year. In that time, she has already made remarkable progress. Originally, I simply wished her to be more confident, daring to say hello to strangers with diverse backgrounds or speaking different languages in a more natural, friendly, and conversational way. Back then, however, many people couldn't accept online English learning. They wondered how the teachers could conduct their classroom activities and help the kids to make progress when everything seemed to be up in the air.

Different parents have different expectations for their children when it comes to English learning. Some want to improve their kids' language proficiency, while others just wish to have a better score on standardized tests. For me, the process matters. I hope that my daughter can actively and happily learn both the English language and about other cultures. The teachers on the VIPKid platform really excel in this area, and I have learned a lot from their teaching activities. Although we are always talking about encouragement in education, personally, I doubt saying "you are great" or "you are awesome" has any positive effect. Cynthia could understand some English when she began to study with VIPKid, because I majored in English, which must have influenced her to some extent. However, her pronunciation was surely not so idiomatic, and it was especially true for her colloquial English.

Sometimes, when she didn't pronounce a word correctly or made a wrong choice, the teachers would encourage her: "Almost right, think about it again." When Cynthia accordingly corrected her pronunciation, the teachers would be full of praise: "Good job, Cynthia." All in all, the teachers encourage kids to think critically and explore new things by themselves. This is important for children and much better than simply writing down the correct answers without knowing why.

Around October 2018, a VIPKid class adviser sent me a message via WeChat, asking me if I was interested in teaching children Chinese. I posted my résumé to Lingo Bus upon her recommendation and received a notice informing me to prepare to meet with VIPKid. My constant company with Cynthia on VIPKid's courses had familiarized me with foreign teachers' teaching styles. In addition, I had benefited a lot from both Lingo Bus teachers' sharing and many online training courses, which enabled me to give lectures smoothly.

The teaching career seemed to open a new door in my life, or another model for me to be a mother. The most obvious manifestation was that I liked to encourage and guide Cynthia to try new things and methods. I used to think it was useless to buy too many toys. Now I have realized I should develop her ability to play with toys. These are all the inspirations I got when I was teaching students in Lingo Bus. Many parents of Lingo Bus students are teachers on the VIPKid platform. They give lectures in the same classrooms where their children's teachers are also having classes. Unsurprisingly, when I show some teaching tools to my students in the class, many children will take out the corresponding items, such as color cards and Dino dolls, in an especially friendly and cooperative manner. They are active and happy in the learning process. Almost every child will take the initiative to greet me and say, "*Nihao, laoshi*" ("Hello, teacher") in a language they are not so familiar with. Although being far away on the other side of the Internet, I am invariably brightened by their smiling faces and enjoy a good feeling all day long.

My classroom never lacks laughter and happiness and is always full of love, for the kids' mothers who must have taught them to speak these greetings before they come to the classroom. How considerate these mothers are!

Gradually, I found there were several kids booking my classes. They belong to two types. Some are very active, and their parents invariably will leave messages, thanking me for making their children happy and vigorous in the 25-minute class. The others are very shy, and their parents will write notes, grateful for my effort in making their kids joyful and relaxed in studying Chinese. My daughter is pretty outgoing, but my husband is very introverted, so thanks to my family background, I am good at guiding and teaching children of these two extreme personalities.

I was particularly moved by a recent encounter. A boy was so shy and timid that he was viewed as dull, slow in reaction, and awkward in communication. After I taught two classes to him, I found the student was only too shy to express himself. In the second class, he was able to introduce his family to me voluntarily. He tried his best to tell me that there were six people in his family, including his elder brother, younger brother, and younger sister. I was really glad that the boy could open his heart to me. His parents left a message expressing their thanks. For the first time, their child voluntarily told them that he was very happy to take Chinese lessons, and he was especially so in introducing his family members to me. But what happened in the third class was completely out of my control. At the beginning, the boy seemed happy to have the lesson. In the last few minutes, he suddenly lay down on the table and didn't respond at all. His parents weren't around, so I had to call him incessantly. Finally, he got up, his face full of tears, and my heart suddenly tightened up. Disregarding the teaching guidelines, I talked with him in English, kept encouraging him, reassured him that he had worked hard and made desirable achievement, for Chinese was very difficult to study, and it was equally hard for me. He slowly stopped crying, telling me that he was too tired and sleepy to go on studying. So I immediately told him to go to bed, assuring him it didn't matter, for the class was over. I also told him no matter what happens in the future, he must tell the teacher in the first place, because teachers will be ready to help him as long as he lets us know where the difficulty lies.

After class, I immediately contacted my VIPKid Fireman and told them what happened. I hoped they would contact the boy's parents and ask them to comfort him. Additionally, I wrote feedback to his parents, encouraging them to tell their son not to be discouraged. It was OK if he didn't want to attend my class. I only wanted to tell him that I loved him. He was really great in learning Chinese. I thought he was just sad for worrying he didn't perform well.

The next day, I still couldn't calm down, because I was not only a teacher, but also a mother. I hope kids will enjoy their studies just as I wish Cynthia will be happy in her learning. I waited anxiously for one day and finally received the feedback from the boy's parents at night. The first sentence was in the tone of the boy, explaining that he was just too tired, for he took part in a competition in the daytime. I was his favorite teacher, and he liked to have classes with me very much. The parents told me that the boy insisted on leaving a message for me personally. They were sorry for failing to consider their child's physical condition. They booked the class too late and did not notice his fatigue at the time. They also thanked me for taking care of and comforting their child. They promised to get everything ready before attending my next class. I was ecstatic when reading this message. The child was fine, and he was willing to continue studying. This is the greatest solace a mother could get.

Many people ask me how I feel as a Lingo Bus teacher. My answer is always that I love it. The class time is flexible, and, hard as it may be, what I do is all out of my love for and dedication to this profession. I think every teacher in Lingo Bus has that same love and dedication. I am being a teacher in the spirit of being a mother. I hope to win love from more children and send happiness and love to more children.

CHAPTER 12
CHANGE AND GROWTH

Our Experience with Lingo Bus
OLIVER'S MOM

I have been teaching on the VIPKid platform for over a year and a half. When I began teaching with VIPKid, I had very little knowledge of China and its culture and knew absolutely zero Chinese. I was very impressed with how well my students, even very young students, spoke English. With two children of my own, I started thinking about their education, and I realized it was an ideal time to start their foreign language learning.

My son, Oliver, was six when he began taking classes with Lingo Bus. He definitely had little to no interest in learning a language, but since Lingo Bus offered a free trial, I figured we had nothing to lose. We also tried a free trial with a competitor company with Spanish language instructions, because my husband and I both speak some

Spanish, and that seemed an easier language to master. Comparing the two services, I loved Lingo Bus more because it was immersive, our teacher was very professional and loving, and the cost was a lot lower. I really did not have much hope or expectation that our son would do very well, though, to be honest. Chinese seems to be such a difficult language.

As we began classes, I would sit with him to help him figure out what the teacher wanted. We watched the preview videos, but it was still completely foreign to both of us, and many times he just parroted the teacher and/or stared blankly. I couldn't help much. He was a ball of nerves, too; he would take the headphone cord and just wrap it over and over around his toes and fingers.

At some point, a door opened in his brain. I think he began to find teachers that he really felt comfortable with and wanted to tease. This was the biggest motivation for his preclass study, actually. I told him, "We have to be ready so that we can trick Steve," and he would diligently practice with me so that he could give sarcastic and witty answers in class. Now he takes great pride in his unique skill. None of his friends are learning any languages yet, and he makes little quizzes for his friends to show off what he knows. My daughter, too, is now taking classes, and her teacher tells me that she often will try to teach her the Chinese colors or some other phrases. I have also learned many Chinese characters and phrases, though I have not had the pronunciation practice, and so my speaking is probably "torture" to the ear of someone in China.

I feel very positively about the impact this has had on both my children. My son does not find school challenging, and I needed a way to stimulate his brain and to challenge him to learn something that he found difficult. I like that he has developed some study habits, and I am thrilled that he has started learning a language early. I hope it will, at the very least, open up another part of the world to him. In just over a week, we plan to make a family trip to China. I know it will be a huge transition for all of us, but I cannot wait to see my children talking with my students in both Chinese and English. It is mind-blowing to me that I teach children in China and that my children speak to teachers across the world. Maybe I'm just old and I never expected this to be possible, but I'm very glad that it is.

We love many things about Lingo Bus. One of those is the teachers, with many of whom we have taken lessons since the beginning. It's amazing to watch old classes and see my son speaking with a horrifying accent and not understanding much of what the teacher said and to compare them with his current ability. I'm sure to an experienced speaker, he has years of work ahead, but I feel very proud of all he has accomplished with the guidance of his teachers. He has figured out how to type using pinyin and often leaves little "love notes" for his teachers for when they enter class.

We also love the lessons and videos. My son really enjoys the Beibei and Aiwen (virtual characters in the Lingo Bus course) interactions. Aiwen is playful and we laugh at how he makes a fool of himself in every situation. The animations are hilarious as well, especially Beibei's scolding looks.

My daughter loves all the songs, and when we prepare for her lesson, she watches them on half speed so she can hear all the words properly. Sometimes in the car, I hear her speaking in slow motion, saying, "If you enjoyed this video, give us a like . . ." She also loves to sing the songs while bouncing on the trampoline. As a family, I feel very good about the investment we have made in Lingo Bus lessons, and I look forward to continuing.

Give Wings to the Dream of "Becoming a Translator"

QIN YUYUE

Qin Yuyue, the mother of a VIPKid student named Liu Jiaxuan, lives in Xi'an, China.

I come from Xi'an city, the ancient capital of China. I am a parent employed in university education. My daughter

Jiaxuan is only nine years old but dreams of becoming a translator in the future.

When my daughter was seven years old, our family lived in the United States for a short time. Being in America for the first time, Jiaxuan couldn't get used to eating bread and other American foods, so she missed Chinese food very much. At that time, my mother, a native from Chongqing city, which is the capital of Chinese cuisine, was so doting on Jiaxuan that she would cook such typical Chongqing dishes as Dandan noodles for her granddaughter at lunch. Whenever Jiaxuan opened the lunch box at school, a group of children would be attracted by the delicious food and bombarded her with various questions:

"Do the Chinese people usually eat noodles? Why are the noodles so slippery? Is Kung Pao Chicken the best Chinese food? Why do the Chinese like drinking hot water?"

In a broken English, Jiaxuan would introduce whatever she knew about Chinese food to her classmates. Sometimes she didn't know how to answer, so she went home to ask me and went to school the next day, with a ready answer for her friends. Thus, at an early age, Jiaxuan had already became a "small window" for her classmates to know and understand China. Meanwhile, Jiaxuan also learned the fact that people from different cultural backgrounds actually have so many differences in the transmission of information, knowledge, and emotion—in other words, she began to experience the fun and significance of cross-cultural communication. So one day, I asked her what she wanted to be when she grew up. At that, she replied, "I want to become a translator!"

Returning to China, Jiaxuan attended a private primary school in Xi'an city. But every time Jiaxuan mentioned friends in the United States to her classmates, she found many of them didn't know the United States very well. At that time, Jiaxuan would draw on her experience in America and explained to them with patience that in the United States, children from all different backgrounds get along with one another and they all like China very much.

It is precisely because of these cultural cognitive differences that Jiaxuan has dreamed of becoming a translator more eagerly and started her journey to her dream, step by step. She studies English even harder than before. Whenever her school has foreign visitors, she will volunteer to become their interpreter. She also takes part in some cultural exchanges between China and foreign countries on a regular basis, acting as a bilingual hostess, interpreting on the site what the Chinese host speaks into English.

I gradually found that the English classes offered by the school couldn't meet Jiaxuan's academic expectations anymore, for she was eager to have contact with more of the world. So I signed her up for VIPKid. On VIPKid, Jiaxuan is even more eager to discuss the differences between Chinese and foreign cultures with teachers. She is also very happy to popularize Chinese traditional culture with them, which makes her closer to her dream day by day.

In addition, Jiaxuan has been lucky enough on VIPKid to make friends with two outstanding teachers who have encouraged her to pursue the dream of becoming a translator.

One is Gina, whose talent is her affinity for conversation and her skill at finding "common topics" with Jiaxuan, which gradually shortens the distance between them imperceptibly. Even I have learned a lot of parenting strategies from Gina. For example, when Gina gave Jiaxuan the first lesson, she did not start with the contents of the regular course but chatted about the books and cartoons that children of Jiaxuan's age were most interested in. Naturally, Jiaxuan was attracted and opened the "chatterbox" immediately with this amiable teacher. Over time, Jiaxuan and Gina have become good friends. She feels more relaxed and can express her views in English more naturally.

In accompanying my daughter to learn English, I come to realize the importance of helping children to guide their interest and to make them have the driving force for learning. After Gina discovered Jiaxuan's interest in the picture books of the *Magic Tree House* series, I not only provided her with more picture books and original cartoon animations, but also acted as the "quality inspector" of her learning materials, which further stimulated Jiaxuan's interest in study. Up to now, Jiaxuan has read nearly a thousand original English picture books and novels and has gradually developed a passion for reading.

Another person who is of great help to Jiaxuan is Teacher JD, a star teacher from Colorado. On VIPKid, Teacher JD has been added by nearly 6,000 parents as their favorite tutor. Indeed, JD is gifted with a magical power: he is able to find out each child's English level accurately and arranges the content of each lesson reasonably, so as to truly implement the principle of teaching students in accordance with their aptitude.

For Jiaxuan, Teacher JD is most helpful in developing her logical analysis and critical thinking. Especially in the free talk time, Jiaxuan didn't know how to give a focused answer at the beginning, thus resulting in irrelevant replies very often. In that case, JD would patiently guide her to find out the contradiction hidden in the problem and seek ways to summarize it. Apart from that, JD often used mind maps to help Jiaxuan grasp the key information of the problem, which effectively improved her abilities in making presentation and summary.

Only later did I learn that JD majored in English literature and was also the captain of the debate team at his university. So he is not only good at critical thinking, but also has outstanding literary attainments. Under Teacher JD's guidance, Jiaxuan fell in love with English writing and began to write her own magic novels. Now Jiaxuan's several mininovels have been widely circulated among her classmates. What is more gratifying to me is that in the process of English writing, I find my daughter fully enjoying herself. Her dedication and persistence have demonstrated her love of writing and even infected me to some extent.

Seeing Jiaxuan become more and more confident, I helped her sign up for an English speech contest. To prepare for the contest, Teacher JD also gave Jiaxuan a lot of encouragement and guidance. He even promised to come to the contest to cheer for her in the finals.

Unfortunately, on the day of the contest, Teacher JD was unable to attend because of his younger brother's birthday, so Jiaxuan and her teacher didn't meet as expected. Although she won the grand prize in the English speech contest finals, Jiaxuan was still a little disappointed.

Teacher JD had the same feeling too, so he wanted to make up for this regret. He contacted us secretly to make a special plan: he would travel thousands of miles to China and celebrate Jiaxuan's next birthday!

On that particular day, I coaxed my daughter to the airport under the pretext that we were going to meet one of my classmates. Jiaxuan readily believed what I said. She was stunned the moment she saw Teacher JD coming out of the airport exit. She remained motionless for two minutes before she ran to embrace the teacher, and both were overjoyed.

Teacher JD accompanied Jiaxuan to celebrate her ninth birthday. Jiaxuan volunteered to take JD to visit the famous scenic spots of Xi'an city, including terracotta warriors, the Mausoleum of the First Qin Emperor, and Huaqing Pool. She also introduced the history of these scenic spots and famous snacks of Xi'an city to the teacher in English. This was a pleasant meeting for both of them.

How time flies! Jiaxuan has studied with VIPKid for two years, during which I have witnessed her great progress. Now, Jiaxuan is about to finish the course of VIPKid Level 6, but she hopes she can learn more slowly, because teacher JD has not yet obtained the Level 7 teaching qualification. So Jiaxuan knows perfectly well that graduating to Level 7 means a farewell to Teacher JD. Understanding how Jiaxuan's feels, Teacher JD promised to obtain the Level 7 teaching qualification, so he could continue studying with her to "go up a level higher."

I was relieved to hear the news. Jiaxuan, my dear little "translator," the wings of your dream have been spread out. I hope my daughter can always smile brightly in the future and fly high into the sky.

My Child is Not a "Prodigy"

TONG YI'S MOM

Tong Yi won a lot of awards in the past two years and was called a prodigy. Here is his family's story with VIPKid.

My son took the high school entrance examination and was admitted to one of the key experimental middle schools in Beijing. In 2017, among all the Chinese pupils in the national delegation, he won the only gold medal of science, technology, and art in the International Exhibition for Young Inventors. In 2018, he participated in the Artec International Camp Exchange Conference and competed with four thousand children from sixty-two countries around the world, winning the gold medal in the English recitation contest. Also in 2018, he led the team to participate in the First Future Space Scholar Contest, winning the championship, and the most popular award on the spot, as well. In the same year, he was designated as the representative of Beijing pupils and delivered a speech at the China Space Day, which was launched by the National Science and Technology Museum. In 2019, he was awarded the copyright (all rights included) of "Dreaming Back to the Tang Dynasty" by the National Copyright Administration. In the eyes of others, my son is a child prodigy. But I don't think so. On the contrary, I believe every child has a 100% chance of creating miracles. It's just that I helped my child successfully find the key to open this miracle gate. In this process, I was lucky to meet such a high-quality education platform as VIPKid, which helped me to become a wise mother who knows how to educate her children scientifically. It also made my son grow up confidently and become more excellent.

As a mother working in a well-known start-up enterprise of Zhongguancun's high-tech industry, I was routinely pretty busy, but I wanted to accompany my son in the meantime so that he could receive the best education. As my son's learning needs became increasingly diversified as he grew up, I came to realize that I should find more learning channels for him. From very early on, my son demonstrated his interest in foreign languages, so I found VIPKid.

The reason I chose VIPKid is, on the one hand, that it adopts an online education mode, thus my son can attend classes anytime and anywhere, which will save me a lot of travel time. On the other hand, the authentic English lessons given by teachers on the VIPKid platform have met my expectation of English teaching—I always believe that children learn English or other languages not only for communication, but also for the cultivation of different thinking modes.

At first, I booked classes for my son very frequently, amounting to three times a week. But in the course of study, I found that he no longer needed to study at my urging. Sometimes I forgot to book classes for him, and he would remind me, "Mom, why haven't you booked any class for me?" So I asked him curiously why he liked VIPKid's classes so much. He told me, "Teacher Justine's class is so interesting. I want to share everything with her. She is already my best friend!" I was astonished by his answer, which is totally different from the effect brought by traditional tutorial classes. I once eavesdropped on what he and his teacher spoke about in class. I was pleasantly surprised by the confidence he demonstrated in the conversation.

At school, Tong Yi's rapid improvement in English proficiency had also attracted the attention of his classmates and their parents. Many parents came to ask me the secret of my son's progress. In my communication with other parents, I found most parents had a severe anxiety over their children's education. Many parents wanted to know why, after taking so many classes in different schools, didn't their children make remarkable progress?

There is only one time for a child to grow up, and the "happy gene" cannot be lost in the child's growing experience. Different from other children whose tutorial classes fill their schedule every week, Tong Yi only attends the English class offered by VIPKid. More often than not, he plays soccer with pupils of grade one or two in the stadium, conducts scientific experiments, and reads extracurricular books in his spare time. The reason why I chose the English course on VIPKid for Tong Yi was that VIPKid had not only greatly improved my son's English proficiency, but more important, it had also cultivated his interest in learning, enabling him to contact with rich extracurricular resources and providing him with a global perspective that would lay a comprehensive foundation for his later participation in international competitions. Therefore, for children's education, blind tutoring is only a waste of their time and energy. It is more conducive to children's growth if parents can choose courses that are really of interest to children and thus beneficial to their future development.

In fact, concerning education, every child is unique and has his own preferences and advantages. As parents, in the process of helping our children to grow up, what we have to do is to help them develop their own personality as much as possible and give full play to their autonomy. Growth is actually a process of revealing the "Matthew effect," looking for breakthroughs in children's personalized development and helping children become open-minded and embracing the world eagerly. If so, all the miracles of the children will.

Looking back on Tong Yi's growth experience, it was not all smooth sailing. No apparent problems were found when he was in the first grade of primary school, but by the second grade, especially in the second half of the semester, I obviously felt that he was very timid and diffident, totally different from what I see in him now. This made me extremely anxious, but after careful analysis, I felt that I couldn't give up. I must actively help my child find his shining points and build up his self-confidence.

Because Tong Yi was good at English, it happened that his school asked him to sign up for CCTV's "Outlook of China" contest. I realized the opportunity had come. Through the study with VIPKid, my son had rapidly improved his English skills. In addition, he also made meticulous preparations. He won first place in Beijing's Early Grades competition and took home the gold medal in the all-Beijing Finals. Meanwhile, he was also awarded the title of "Excellent English-Chinese Bilingual Host." This time, I invited Mr. Gao, the principal of their school, to present the award to him in person, in the hope of inspiring a stronger sense of honor in him, because the principal has great influence in children's minds. Even now, I am still very grateful to Mr. Gao, for I remember clearly how excited my son was when he told me after school that the principal awarded the prize to him personally.

Later, my son won one honor after another. In the same year, Tong Yi won the first prize in the national finals and various awards in DI global competitions. His self-confidence was building up. With the support of teachers from the science and technology community, he has won more awards and become the student with the most prizes in his school. Therefore, my conclusion from Tong Yi's experience is: Methodically training children in the fields where they are good at, and letting children exert their own advantages to create miracles, can subjectively change their subconscious self-cognition, help them build self-confidence, and make them grow in a healthier and all-around way.

Finally, as a mother, I fully understand parents' expectations for their children, but when facing educational anxiety, we should respect our children, give our children appropriate expectations, and explore the most suitable educational methods for our children in the meanwhile. Fortunately, Tong Yi is doing well at this stage. Here I would like to thank VIPKid. Its existence makes me get twice the results with half the effort on the way to educate my son and gives me more confidence in his future growth.

Embrace a Bigger World

WANG ZHUZHU

Wang Zhuzhu(Jenny), Ma Qixuan (Oliver)'s mother, records a story about her son's growth with VIPKid.

I am the mother of a nine-year-old child in Xi'an city, China. My son's English name is Oliver. He is an enthusiastic, cheerful, and humorous child, often telling jokes to amuse friends around him. He has many interests and hobbies. For example, he likes playing basketball and is such an enthusiastic player that he seems infinitely full of energy. But he also has a quiet side. He likes playing the piano, not to pass requirements, but with a genuine interest in the world of music.

I am studying for a doctorate degree in computer science, and Oliver's father is a university teacher. As parents, we have a very simple wish for our son's future development: We hope that he can love life and receive a good education. We also hope that he can foster outstanding abilities so as to choose the life he desires and embrace a bigger world.

Two years ago, Oliver went to an American school to study as an exchange student. We took Oliver to Salt Lake City and lived there for seven months. Oliver was admitted by a local primary school in March and studied in the second semester of the first grade, which turned out to be a great challenge for him.

Before going to the United States, Oliver could not write a single word, and his English skills were poor. He only learned such simple words as "hello" and "thank you" in kindergarten. Therefore, my husband and I became anxious, worrying whether our son could adapt easily to American primary schools.

However, this didn't seem to have caused great obstacles for him and his study in the United States. On the contrary, his English had improved rapidly.

The credit might be given to the teaching methods in the United States. Different from the teaching in China, the educational context in the United States is relatively relaxed, and Oliver was very willing to integrate into the local students' way of living and learning. His head teacher also gave him a lot of extra help and encouragement. For example, a teacher in his class would be specifically responsible for Oliver's teaching, using some gestures to help him understand and communicate with other people.

After only two months, Oliver could understand his teachers' lessons, and he also made friends with many American children. At first, we would explain what the rules were for American children to play games in order to help Oliver to join them. Oliver was silent in the beginning, but after a period of time, we could hear through the window that his laughter was all over the yard. Even when my husband and I were still surprised by this change, Oliver himself was already enjoying a foreign life.

Our time in the United States passed quickly. Many interesting things occurred during our stay, but Oliver brought to us still more surprises. What impressed me most was an anecdote that happened in Oliver's second semester. Once the school invited students' parents to a composition meeting, in which the teacher called us to the school and praised Oliver's composition.

"Oliver is so good at exaggeration and metaphorical rhetoric that I can't help laughing whenever reading his composition!"

But the teacher became even more surprised when we told him that Oliver only studied in the United States for a few months.

"I always thought he was a child raised in the United States! He is so skillful at both written and spoken English."

Of course, hearing the teacher's positive feedback about Oliver, my husband and I were overjoyed. Finally, by the time we left the United States, Oliver had exceeded 90% of his American classmates in terms of academic performance.

Since my husband's working period was going to expire very soon, we had to consider the arrangements after returning home.

Before returning home, what my husband and I worried about most was that once Oliver left an English-speaking environment, he might forget what he had learned about the English language. So before going back, I asked around in order to find some desirable English lessons in China. Some friends who had similar foreign study tours recommended VIPKid and told me it was a famous online learning platform. Oliver also said that some teachers in his school were teaching on the VIPKid platform. Therefore, on the return flight, I signed

Oliver up for VIPKid, so in the first week after returning home he could start his English learning on the platform.

Oliver entered the affiliated primary school of a university and continued to study in the second grade. The strict teaching method of Chinese teachers is quite different from that of their American counterparts. Sometimes a set of examination questions have to be repeated many times, so Oliver missed the learning environment in America very much.

We didn't want to add more study burden on Oliver, so all of his extracurricular classes were his favorites and chosen at his own will. This included basketball, piano, and VIPKid. For him, VIPKid is an extra reward besides his regular school study, and the learning is a pleasure. In addition to the major courses with VIPKid, Oliver also enjoys such minor courses as "Foreign Teachers Show You the World" and "Word Fighting," in which he has learned a lot of authentic English expressions. Every teacher on the VIPKid platform has a unique personality, and many teachers have become good friends with Oliver. Among them, three teachers have impressed me deeply. One of them is Chris AH, a very young and funny male teacher, with the same personality as Oliver. They always share many interesting trivial things in life. For example, Oliver had just raised a cute little turtle, so he couldn't wait to tell this new pet's name to his teacher-buddy. On the other hand, Chris AH loves coffee very much, and he often shares his coffee experience with Oliver. Influenced by him, Oliver gets used to jotting down any information about coffee whenever possible, so he can subsequently discuss this newly acquired knowledge with Chris AH in class.

In the course of study, they are often heard bursting into loud laughter. Oliver is so attached to his teacher-buddy that he frequently tells him some interesting things that happened in our family.

Unlike humorous Chris AH, Oliver's Timothy O is an insightful teacher. I admire him very much because he often draws on his years of teaching experience to help Oliver solve difficulties arising in the process of learning, and, more important, he helps my son form a habit of considering things in the long run, so that Oliver can face the challenges in the future by himself more calmly.

Oliver also likes a teacher named Lisa ME, who has nineteen years of teaching experience and has mastered a unique teaching style, which brings a special affinity to her. In Oliver's words, "Lisa ME is like a grandmother." In the class, when she finds Oliver is inaccurate in pronunciation or knowledge points, Lisa ME will be very patient to make demonstrations, explaining again and again until Oliver fully masters the relevant skills.

These teachers, each with different personalities and teaching methods, have all given Oliver a rich and varied learning experience.

The travel experience in a foreign country has given Oliver his first taste of the bigger world, and VIPKid's online learning experience allows Oliver to continue his communication with the world and also enables him to have a broader understanding of the world. With regard to our son's growth in the future, we hope he can learn English better, accept different cultures, and embrace the world. Meanwhile, we hope that Oliver can face the future with an open mind and in an earnest manner, so he can grow up into a man with an international vision who knows what he should do for his home and mother-land in the future.

EPILOGUE
By Cindy Mi, Founder and CEO of VIPKid

As I looked out into a sea of orange, the passion and excitement in the room was palpable. Over 500 teachers came from across the United States and Canada to gather together in the historic Hilton Chicago for one purpose: to engage with and celebrate their fellow teachers. I was there to express my gratitude to them, to meet them and hear their stories, and to share with them my vision for the global classroom that they are helping to build.

At that same event in March 2019, we announced that we had reached 600,000 students and 70,000 teachers on our platform, so the mood at our Journey conference was celebratory and optimistic. I saw teachers forging connections with one another and with our US and Beijing staff. Our ever-popular Dino mascot was making its way around the conference, taking photos with groups of cheering teachers. The energy was contagious, but I took a quiet moment to observe and take in the amazement of how much VIPKid had grown and the incredible community that supported it.

Soon, I was standing in front of a line of teachers who were waiting to take photos with me and share their own VIPKid stories. For many, VIPKid had become a way of life and something they look forward to every day, just as our students do. The lives of so many have been changed. Today, one in two Chinese children who learn English online are part of the VIPKid community. Though we had to struggle to achieve success, I see now that the dream that I had years ago is becoming a reality.

Review

In 2013, I founded VIPKid based on the simple idea that every child deserves a global classroom with high-quality teachers and content in a personalized learning style. At the time, there were only 27,000 English teachers from North America living in China. Meanwhile, over 15 million babies are born in China every year. There simply was not enough supply to meet the demand for English language learning at brick-and-mortar tutoring centers or private in-person tutoring. Increasingly globalized and educated parents in China were seeking ways to give their children an equally global education from the best teachers in the world. Years ago, they would have had to settle for finding a tutor living in China and potentially give up their evenings or weekend mornings to drive their kids to these classes. With technology advancing so rapidly, I realized there was a better way.

When I was seventeen, I started a brick-and-mortar tutoring company, ABC English, with my uncle. It was there that I developed my passion for education and saw that parents' and students' needs were not being met fully. At ABC English, I was involved in every aspect of the business, from curriculum development to teacher recruitment and business operations. It was on this foundation that I felt prepared to start a new technology-based startup. However, not everyone shared my vision, and many questioned whether it was even possible. What I had in mind would not be easy. Fortunately, my passion and optimism won out, and I forged ahead with drafting a business plan and recruiting cofounders.

On October 18th, 2013, VIPKid was officially founded. At that time, we only had a few work desks in Zhongguancun Technology Park, a technology hub in Beijing known as "China's Silicon Valley," where I worked with a small team to develop and pilot our technology and business model. Those days were spent working around the clock, with no promise of a payoff in the end. It took nearly a year and a half to finalize the product and ensure it was effective. Quality was our guiding principle. We stayed focused on what we believed VIPKid could become and spoke to as many potential partners and investors as we could for support and guidance.

Even with the technology formalized, we still needed

students and teachers. Many parents were skeptical that their children could actually learn online, and we struggled to sign up our first students. My cofounder, Jessie Chen, had a young son, and he became our first student. The children of some of our early investors also volunteered to try it out. With a few students, next we needed the high-quality English teachers that Chinese parents sought.

To find the best teachers, I traveled to the US to personally recruit them myself. From coast to coast, I networked and spoke with teachers, eventually convincing ten to sign up on the platform. It wasn't easy, and they took a leap of faith by joining us. Then, the moment of truth: Would the student-teacher relationship flourish online? Would the student truly learn and would the teacher be able to connect with the student effectively?

It was clear from the first class that the answer was yes. The student was engaged and joyful, and the teacher was able to use TPR to communicate her lessons. The teacher even took out a ukulele and sang songs with the student. When I observed this first class, my spirits were lifted, and I was recommitted to growing VIPKid as much as I could. Soon, we had one hundred students, mostly through word of mouth. Parents loved the convenience of having a tutor virtually visit their home every night and were impressed with the results. Our student numbers began rising so fast that at one point, everyone in our office had to help with sales and on-boarding.

Meanwhile, teachers overseas were starting to discover VIPKid and referring their neighbors, friends, and fellow teachers. They loved having their own business and making extra money each month, all while doing what they were passionate about—teaching! By 2016, we had 7,000 teachers. In 2017, we grew to 100,000 students. Our growth was so rapid, and our offices in Beijing were working around the clock to ensure that our technology could scale.

We knew then our next step was to open a US-based office to help nurture and support the growing teacher community. We were surprised at how teachers had formed their own unique community—both offline and online. Teacher meetups, where one teacher hosts a group of teachers at their home or in their community to bond over their experiences and share tips, were popping up all over the US. We recognized the power of this community and wanted to strengthen it. We were inspired by these meetups, and that's when the idea for the Journey conference was born.

Our first Journey conference was in Salt Lake City in March of 2018. We had a small team of US staff who worked with teacher volunteers to design the event. We didn't know what to expect and were overwhelmed as teachers flooded into the Little America Hotel, dressed in orange and in high spirits. When Dino made a surprise appearance, the teachers went wild! The whole day was so special and focused on celebrating each teacher's own journey. I felt uplifted and energized to keep working to build a global classroom even bigger than I had imagined before. I realized then that the teacher community is so important to the future of VIPKid.

Back home in China, we knew we could do more to expand access to this global classroom. In partnership with the Jack Ma Foundation, we embarked on a mission to give back to society by bringing education and technology to rural areas in China that have limited Internet connectivity. I first visited the villages in rural areas of Sichuan Province and Yunnan Province of China. What started as a humble project now aims to bring high-quality education to over ten thousand schools in rural China.

It's been almost six years since I founded VIPKid, and our mission has not changed. Every day, we are working tirelessly to bring a high-quality personalized education to children across the world. We still have miles to go on our journey, but I can promise you that we will not stop.

Giving thanks

Of course, the journey to a truly global classroom cannot be undertaken alone, and there are many people who deserve my endless gratitude and thanks.

First, thank you to the parents who put their trust in us every day. We care deeply about helping your child succeed and will always be focused on improving VIPKid to help them learn better. I had many sleepless nights thinking about how I can ensure that our hundreds of thousands of students are receiving the best education in the world. I recognize that your child's education is one of your top priorities in life, and I thank you for giving us the opportunity to contribute to it.

Additionally, we would not be where we are today without the teachers. They are the lifeblood of VIPKid. To every teacher who has taken a chance on teaching with a Chinese company that they may have never heard of before, I thank you for believing in us and our mission. Whenever I visit the US or Canada, I try to meet as many teachers as I can. I love hearing stories from them about how VIPKid has changed their lives; whether they can now stay home with their young children or have fallen in love with teaching young kids across the world, they are truly inspiring to me. I can feel their passion and dedication to teaching when I'm in the room with them.

Alongside me in the office every day are thousands of employees who are just as dedicated to fulfilling our mission as I am. When people ask me what I am most proud of, I always answer, "My team!" VIPKid would not be the leader in K-12 education without the hours of work put into it by our employees. Even with a thousand words, I cannot adequately express my gratitude and admiration for our team, and I know we will look back and remember this time as one of hard work but also great purpose. Thank you for being with VIPKid all the way!

To our partners and investors who are diving head-first into building the global classroom with us, thank you for your support and contributions. They are truly leaders in their field and have provided their expertise and guidance to us along the way, making a long-lasting impact on our curricula and service. We will also be forever grateful to our investors, who believed in us from the beginning and who have given us invaluable guidance.

With the support of our parents, teachers, employees, partners, and investors, the future of VIPKid is brighter beyond our imaginations.

Looking to the Future

As I travel around the world and share the story of VIPKid, I'm often asked about the future of education and technology. My vision for the future is a world in which technology enables us to connect with others, improve education outcomes, and transform the way that teachers can make a living.

I envision a future where the global classroom will connect every child to the best teachers across the globe. Imagine a world where students from Chicago take a virtual field trip to the pyramids in Egypt with students from China, learning from a teacher who is an expert in her field. In this future, education will become borderless, and geographical barriers will be broken. The classroom will no longer be tied down by physical constraints; it will be a true classroom in the cloud. I firmly believe that when children can learn about the world around them by meeting other children from across the world and experiencing other cultures (even virtually), they are on the path to becoming global citizens.

I envision a future where a child will be able to learn in a personalized way at his or her own pace and learning style. Technology will enable us to improve the classroom in many ways. We can build "study buddies" for students that can guide them through their homework and provide feedback in real time. Technology allows us to understand where a student is struggling and can relay that information to their parents and teachers. It's a tool that we can utilize to make learning more fun. For example, when a student is able to see dinosaurs come to life through augmented reality, their curiosity is immediately sparked. Through bringing technology into the classroom, children will be able to reach their potential and become lifelong learners.

I envision a future where technology enables more equity in education. We are now able to livestream teachers into classrooms in the most rural parts of the world. VIPKid today provides our service for free to rural schools in China, which would otherwise be very costly. We now offer free high-quality English language learning taught in real time online by teachers in the US and China to 1,000 schools in rural China, and we have plans to help more children there within just a few years. As we expand our ability to provide rural schools in China with English learning opportunities, we are beginning to explore ways to provide underserved children in the US.

I wish that I had had a global classroom like this when I was a child. Every child deserves a global classroom that is accessible, affordable, and personalized. The reason why humans are able to create a better world is that we have education. Every tiny change and innovation in education may have far-reaching significance for human development. I would encourage you to join the VIPKid mission so that we can work together to inspire and empower every child for the future.

FURTHER READING

Cindy Mi, Founder and CEO of VIPKid, talks about what separates VIPKid from the competition
https://www.youtube.com/watch?v=BC-CFiFwigA

Interview with the Chinese entrepreneur revolutionizing education
https://www.youtube.com/watch?v=Krdn_RvTQsM

Cindy Mi on Building VIPKid, the World's Largest English Learning Platform for Children
https://www.youtube.com/watch?v=eJVTnVVJtgY&t=15s

How VIPKid CEO Made Education a Universal Language
https://www.fastcompany.com/40525523/how-VIPKid -ceo-cindy-mi-made-education-a-universal-language

Cindy Mi and Qi Lu Share Advice for Entrepreneurs Building Global Companies https://blog.ycombinator

.com/cindy-mi-and-qi-lu-share-advice-for
-entrepreneurs-building-global-companies/

How VIPKid Became the Biggest Unicorn in China's
   Fiercely Competitive EdTech Market
https://medium.com/@EdtechChina/how-VIPKid
   -became-the-biggest-unicorn-in-chinas-fiercely
   -competitive-edtech-market-9969602705bd

National Center for Education Statistics
https://nces.ed.gov/fastfacts/display.asp?id=372

American students have fun learning Chinese with
   Lingo Bus
http://www.chinanews.com/business/2018/04–08/8485552
   .shtml